Environmental Policy in Search of New Instruments

EUROPEAN SCIENCE FOUNDATION

This series arises from the work of the ESF Scientific Programme on Environment, Science and Society: an economic and institutional analysis (ESS).

The Scientific Steering Committee of the ESS Programme is composed as follows:

Environmental Policy in Search of New Instruments

Edited by

BRUNO DENTE

Kluwer Academic Publishers

Dordrecht / Boston / London

Library of Congress Cataloging-in-Publication Data

Environmental policy in search of new instruments / edited by Bruno Dente.
p. cm. -- (Environment, science, and society ; v. 3)
ISBN 0-7923-2949-X (alk. paper)
1. Environmental policy. I. Dente, Bruno. II. Series.
GE170.E5767 1994
363.7'05--dc20 94-21240

ISBN 0-7923-2949-X

Published by Kluwer Academic Publishers,
P.O. Box 17, 3300 AA Dordrecht, The Netherlands.

Kluwer Academic Publishers incorporates
the publishing programmes of
D. Reidel, Martinus Nijhoff, Dr W. Junk and MTP Press.

Sold and distributed in the U.S.A. and Canada
by Kluwer Academic Publishers,
101 Philip Drive, Norwell, MA 02061, U.S.A.

In all other countries, sold and distributed
by Kluwer Academic Publishers Group,
P.O. Box 322, 3300 AH Dordrecht, The Netherlands.

Printed on acid-free paper

Printed in the Netherlands.

The *European Science Foundation* is an association of its 55 member research councils, academies, and institutions devoted to basic scientific research in 20 countries. The ESF assists its Member Organisations in two main ways: by bringing scientists together in its Scientific Programmes, Networks and European Research Conferences, to work on topics of common concern; and through the joint study of issues of strategic importance in European science policy.

The scientific work sponsored by ESF includes basic research in the natural and technical sciences, the medical and biosciences, the humanities and social sciences.

The ESF maintains close relations with other scientific institutions within and outside Europe. By its activities, ESF adds value by cooperation and coordination across national frontiers and endeavours, offers expert scientific advice on strategic issues, and provides the European forum for fundamental science.

This volume arises from the work of the ESF Scientific Programme on Environment, Science and Society: an economic and institutional analysis (ESS). This programme stimulated a number of successful collaborating networks between economists, policy analysts, lawyers and natural scientists across Europe.

Further information on ESF activities can be obtained from:

European Science Foundation
1, quai Lezay-Marnésia
F-67080 Strasbourg Cedex
France

Tel. (+33) 88 76 71 00
Fax (+33) 88 37 05 32

Contents

1. Introduction: The Globalization of Environmental Policy and the Search for New Instruments

BRUNO DENTE
Istituto per la Ricerca Sociale, Milan, Italy

1. Environmental Policy in Transition: The Challenge of Globalization

It is probably unfair, and even somewhat ironic, to say that environmental policy is in crisis. Thousands of government officials at all levels, millions of citizens, scores of interest groups and hundreds of scholars of many different disciplines spend a large part of their time worrying more about the environment of the whole planet than they do of their backyard. Industrialists and economic interests are deeply concerned about the limits that a growing body of legislation is imposing on their operations, and still they are not able to reverse the trend. Environmental parties are probably less important today than they were a few years ago in many countries, but this is more than compensated for by the fact that the traditional parties have incorporated many of their claims. The U.S. presidential election of 1992 was fought perhaps as much on environmental issues as it was on economic and fiscal issues. As far as public policy is concerned, the creation of new administrative departments and technical agencies is clear proof of the fact that the priority of environment in governmental agenda is very high indeed. And, last but not least, the global dimension of environmental policy is the basis of a growing body of international treaties and agreements: the UNCED meeting in Rio de Janeiro was certainly one of the most advertised international meetings of the last decade. If the results were somewhat disappointing, there is no doubt that the future will see more of this kind of event.

The reasons for the rise in environmental interest are obvious. Apart from the mere pressure of the problem (large enough of course) the protection of the environment seems to provide a shared ideology to political systems which have experienced the traumatic fall of communism, a new emphasis for public bureaucracies in search of a mission after the – at least partial – demise of the Keynesian welfare state in a period in which the private consumption seems to have reached, in rich countries, a limit very difficult to overcome. Advances in knowledge emphasize the bad state in which our planet is and will probably be in

Bruno Dente (ed), Environmental Policy in Search of New Instruments, 1–20.

the foreseeable future, legitimizing (actually requiring) new efforts in prevention and cleaning up.

So, where's the crisis?

Maybe 'crisis' is too strong a word, but a careful observer can certainly perceive a certain unease if not a true disappointment. Certainly many factors account for this situation.

In the first place, some reaction to the honeymoon of the past was to be expected. After the spectacular successes of the first years in curbing the pollution, the results of subsequent efforts have been somewhat small and slow in coming. Is this due to the iron law of diminishing returns, as has been suggested by some, or the fact that the new environmental problems are much more complex and difficult to tackle than, say, sulphur dioxide in big cities? The answer is immaterial, but the fact remains that opening new fronts every day has the effect of concealing from the people the simple truth that certain promises are difficult to fulfil.

In the second place, the difficult of the present economic situation represents an obstacle not easy to overcome. In the affluent 1980's there was money enough in the public as well as in the private sector to invest in pollution control and cleaning up. Today, with the recession hitting the western economies hard, there are increasing calls for austerity in spending and decision makers will naturally start to wonder if, after all, stringent environmental protection is a luxury we cannot no longer afford. To shift the burden onto future generations seems not a bad idea to a population fearing a drop in their standard of living.

In the third place, the discovery of the global character of environmental problems, the emphasis on the planet as a whole (implicit in the greenhouse effect and the ozone layer depletion debates), of course contributes to the high priority of the international dimension, but also poses tremendous problems. In a way the escalation of environmental concern, as a strategy for gaining public acceptance, is self-defeating, because it runs the risk of devaluating what it is possible to do, here and now, at the local and the regional level, in order to protect ecosystems. The emphasis on the global challenge increases the temptation to free-ride, and shifts the responsibility from the individual polluter to the international decision makers, whoever they may be.

There are probably other reasons for the uneasiness we are experiencing, but our immediate concern is for the public policy itself and here, we believe, there is a crisis of a sort, probably not fully understood and yet more dangerous than it appears.

First of all one has to remember that governmental intervention in the field of environmental protection is a quite recent phenomenon. In some European countries certain aspects of the ecosystem have not been as yet legislated upon and, even where the scope of the intervention is more or less comprehensive, the experience accumulated remains too insubstantial to provide a clear picture of the effectiveness of the activities in the long run. Even accounting for the huge progress made in recent years, at least as far as public alarm for the quality of the environment is concerned, the time which has elapsed since the inception of the policy is too short to have created enough experience and develop a true learning mechanism is several fields.

But the fact that the policy is relatively young has another important consequence. Not unlike other policies in their embryonic phase, environmental policy, in order to get recognition and increase its own importance, has until very recently stressed its own 'special character'. In order to protect itself from other policies, and to expand – successfully, as we have seen – its scope and its resources, environmental policy has tried hard to show that its own concerns are very different from those of, say, health policy or land use planning.

This is not a strange occurrence. The process of establishing a new permanent policy field will usually follows this pattern. On one hand stressing its 'special character' has the effect of reassuring other policy communities that the newcomers do not have the intention of trespassing into other, well established, turfs. On the other hand insisting on the unique nature of the problems, on the specialist skills needed in order to get results, and the distinctive characteristics of the instruments to be employed makes a claim for obtaining direct access to the relevant resources. So far as national bureaucracies are concerned, the symbolic – but at the same time very practical – achievement that crowns these efforts is the creation of departments or ministries of the environment, which are headed by a, frequently senior, member of the Cabinet, who in their post become responsible for the dynamic adaptation of the intervention to the changing conditions. Such departments do of course participate in the budgetary game in order to get the necessary financial resources, and develop their own network of contacts, mainly with interest groups, in order to provide themselves with a stable and strong constituency.

However not only is the first phase this predictable, but also the steps which follow may be seen to accord with a definite pattern. After the special nature of the problem has been asserted, and become widely recognized, the focus of the policy actors will tend to shift towards the need for integration, or at least coordination, with other policy fields. Of course the promoters of the "new" policy – that has by now become an established one – will hold the position that such a coordination is synonymous with the superiority of their policy *vis-à-vis* other policies. Integration and coordination will be equated in their attempts, with the recognition of the overarching importance of their problems, and if they succeed, of themselves. It is to be expected that the other policy activists will counteract this move, claiming the superior nature of their own problems or, at least, their problem's special character. This kind of situation has been experienced quite often in the history of public bureaucracies, and the outcomes can differ according to the circumstances. In most cases the game reaches an equilibrium of a sort, based on the mutual recognition of the spheres of influence and possibly, the institutionalisation of cross veto powers. But, in some cases, the struggle can go on in cycles until someone actually wins and achieves a special status in the policy arena (think of monetary policy today), at least until something changes in the outside world. The contradiction between on the one hand claims of a special character, built into the embryonic phase of a new policy, and on the other hand the push towards coordination and integration, which characterizes the second, more imperialistic phase, is by no means a feature specific to environmental policy.

Environmental policy, however, has a peculiar aspect, at least today. Its global character, being embedded in the very nature of environmental problems, has apparently emerged, thanks also to growing international pressure on individual countries, at a rather early stage, even before there has been enough resource accumulation in order to fight and win the interdepartmental game.

We have already suggested that one of the challenges posed by the globalization of the environmental issue is the apparent contradiction posed by the need to legitimize environmental policy in terms of results. We can illustrate this by indicating how the specific crisis that environmental policy will have to face in the very short term is, in many countries, a dilemma between either retaining its specialist character, i.e., underplaying the global aspect, or going for a more comprehensive approach and perhaps, running the risk of losing, at least partly, the status already attained.

Of course this choice is conditional upon the resources that the different national environmental policies have been able to mobilize and accumulate in the recent past. Of course some countries are in a better position to choose the second strategy than others. The point is, however, that, as we have already suggested, playing the global game is not without its dangers. If the quality of the environment in a European country depends upon Brazil doing something about the rain forest, and the USA curbing the greenhouse gases emission, then the incentive to act at the local level would be rather low. In other words environmental policy has seen a rapid escalation of the problems without experiencing a corresponding increase in its own ability to cope with them. Put differently, one can suggest that the comprehensiveness of the issue has outgrown the level of institutionalisation of the policy itself.

The demand for new instruments stems from this conflict. The main policy actors (foremost governments) need new ways of addressing both old and new problems, knowing very well that in the short run they will be judged on the basis of their ability to provide satisfactory answers to issues that, like the proverbial *apprentis sorcier*, they have contributed to the creation of.

Before discussing what a policy instrument is, one has to remember that another symptom of the immaturity of the policy is the rather naive way in which the whole issue on instruments is defined and debated. As Peter Jan Klok shows in his chapter, the least one can say about the current classification of environmental policy instruments, including the omnipresent distinction between regulative and financial instruments, is that it is ill-formulated. Here again one may suspect that we are looking at the tip of the iceberg. Fundamentally the current debate on environmental public policy often seems to ignore the results of a long tradition of scholarship in public administration and policy analysis, with the danger of inventing the wheel over the over again. One good example could be the proposal to introduce into this field the same type of meta-decisions that have already failed in budgetary policy (like PPBS).

In short, I believe that, in order to find new ways of coping with both new and old environmental problems one has to start from the good old theory and methodology of public policy analysis, in order to identify the alternatives that are open to policy makers. Environmental policy is, first and foremost, a public policy, and there is no reason to believe that the rules that apply to it shall be very different from those which evolved, over a longer period of time, in other policy fields.

So, to proceed in a scholastic way, what is a policy? The standard definition tells us that it is a set of actions connected to the solution of a policy problem, i.e., of an unrealised value, need or opportunity which, once identified, may be attained through public action (Dunn, 1981). According to the tradition of policy studies, there is no need to restrict the notion of public policy to the conscious attempts by governmental institutions to tackle specific issues. Market mechanisms or simple mutual adaptation by the relevant actors can, under given conditions, work equally well in solving the problem. But even when the latter stems from a market failure – as it is in the overwhelming majority of cases in the environmental field – this does not automatically imply the authoritative intervention of a governmental body, in the form of regulative action or of a fiscal incentive. According to the level of complexity of the policy field other roads can be followed and can prove equally, if not more, effective. This does not imply simply the raising of public awareness generally or the use of 'persuasive' instruments. One example is the collaborative agreement on energy conservation reached by the environmentalist groups and the energy industry in California (Mazmanian, 1991). Or, better still, one can quote the agreement reached in the Netherlands between environmentalists and consumers to phase out PVC in certain packaging applications (Tummers, 1991). Or again one can think of the role of the experts – the so-called 'epistemic community' – in bringing about the international agreements in order to protect the Mediterranean (Haas, 1990). All these examples show how there are indeed more options than the current philosophy of the environmentalist specialist community seems to perceive.

As I have already suggested, the oversimplified quality of a lot of environmentalist discourse – at least as far as the public policy dimension is concerned – results less from a lack of creativity and more from a lack of method, and of a firm theoretical foundation. This failure leads to non-cumulative knowledge, and eventually to a growing confusion in understanding what the real alternatives in policy making are.

Drawing upon the chapter by P.J. Klok, and the vast body of literature he summarizes, one can say that the effects of a policy (in terms of outcomes, impacts and outputs) can be explained by:

1. the nature and the number of actors;
2. the nature and amount of resources they command and exchange between themselves;
3. the way in which they interact.

All these variables (the actors, the resources and the patterns of interaction) can be further analyzed, classified and measured. All of them in general, and most of them in any given empirical setting, can be altered by some participant actor through conscious manipulation. We can therefore define a policy instrument as the way in which an actor (say, a public agency) can modify one or more elements of a policy game in order to get a desired result. For instance one can think of modifying the set of actors, by introducing new players (or, less likely, by forcing someone out of the picture), one can vary the resources available to the participants (increasing or decreasing the rights or obligations, the political consensus, the money, the information, the knowledge and/or the time), one can try to alter the way in which one or all of the actors relate to any one or even all the others (for instance, by unilaterally withdrawing from the negotiation process, or by creating a common understanding through which a positive problem-solving attitude can be developed).

This is of course a very rough and crude picture of how a policy game can be played (for instance one should at least discuss the thorny issues of problem definition and structuring, of the role played by the stake of the game, of the strategies that the different players can adopt, etc.), but even from this bare sketch it is immediately apparent how environmental policy, being young, is still a rather simple policy. In reality the resources utilized have, to a disproportionately large extent, been normative and authoritative in nature, underplaying the other types, with the only possible exception being that of scientific knowledge. The patterns of interaction, therefore, are mainly of the authoritarian or hierarchical kind, and the bargaining inevitably creeping into the processes is very often regarded with suspicion, if not as being downright dangerous and non-legitimate. But what seems even more important is the fact that, in the vast majority of cases the number of actors is rather small, in terms of the types of interest represented. At least in the European context, it is usual in policy implementation – and often also in policy formulation – that the policy community (or the issue network) is composed solely of the technical bureaucracies, the polluters and/or their representatives, and the experts. Only in a major case do the environmental groups have a relevant role, and even less frequent is the intervention of the politicians, of the general public, of the courts and of the representatives of other interests and/or other policy fields.

One of the consequences of this is that, because of the concomitant dominance of normative resources and authoritarian patterns of interaction, when the game becomes really complex – for instance, because the sudden increase in social alarm creates the conditions for an increase in the number of actors involves – the outcomes are totally unpredictable, if there are any outcomes at all. Here again the globalization of the issue, and in particular the growing consciousness of the fact that the alteration of one natural element has or can have far reaching consequences in many different ecosystems, seems to be in conflict with the way in which the policy making process is usually structured. The outcome, paradoxically, is very often precisely that when the social alarm is higher, because of the recognized global nature of the problem, and when, consequently, the advocates of environmental protection should be more powerful in forcing some sort of action (at least because the political consensus they command is larger), they are at loss in playing the game, because the instruments they can think of using are too narrow in scope or too simple in coping with complex situations.

2. Looking for New Instruments

It is in the direction of enlarging the alternatives open to decision makers in the field of environmental policy that we have directed out efforts here. Since the beginning of our work we were conscious of the need to look for 'new ways' and 'new ideas' for an environmental policy. The challenges for globalization, of a more comprehensive – and preventative rather than remedial – approach to the protection of the environment and also the awareness of the dangers involved in over-regulation pushed us in this direction. This collection of essays represents the result of our common work. Obviously it has no pretence whatsoever of providing a full picture of what can be done and how it can be done, but is intended only to enlarge the range of tools available to the environmental policy actors. What we hope to have achieved is the provision of some new insight into possible courses of action, and some methodological indications about how to proceed and where to look.

As I have already anticipated, the first suggestion we make is to direct analysis squarely towards the relevant public administration and public policy domain. Peter Jan Klok's chapter achieves exactly this, building an approach to the question of policy instruments which goes beyond creating a mere inventory of existing experience, and which avoids the traps of a rigid classifications system, which would run the risk of becoming quickly outdated by the rise of new instruments. In his chapter he confronts the issues posed by there being a limited amount of resources available to the actors, discusses the basic tactics that can be used in order to influence the behaviour of the target groups and he proposes a notion of the policy instrument, more complex than, but theoretically compatible with, the definition given above. Furthermore he stresses a very important issue, seldom raised in the debate. Every proposal to employ a new tool implies not only certain assumptions and hypotheses about the factors that influence human behaviour, but it is also strategical in nature. Except where the administrative agencies concerned (or, for that matter, the environmental groups) are involved solely in the physical removal of the pollution, as is sometimes the case, they are taking actions of a strategic nature, i.e., they are in the position of having to anticipate the reaction of the target groups. Unless these assumptions are made explicitly there is always the risk of designing policy instruments that are not able to account for real world actors, and of overestimating, for instance, the willingness of the different polluters to comply with a regulation or believing that a market for trading pollution permits will automatically appear. This needs to be taken fully into account. The strategic aspect is also stressed in the two following chapters with reference to the process of adoption of new tools.

As Corinne Larrue suggests, there are a lot of good reasons why economic instruments are not used more widely. The obstacles can be procedural, institutional or structural in nature, but all together they have been able to limit effectively the adoption and/or the implementation of proposals such as user charges or marketable permits. The fallacy of the reasoning underlying many complaints against this 'incomprehensible inertia' is a very usual one among non-specialists, i.e., the idea that 'the government' is a single, rational actor, with perfect information and no costs in decision or enforcement. As Corinne Larrue reminds us, the governmental agencies responsible for the formulation and implementation of public interventions are several, different and very often in conflict with one another. Even more important, they are composed of men and women having diverse skills, ideas and interests. Not only has one to take all these elements into account when putting forward a new proposal, but one has to be conscious that during the process of arriving at a decision – but often also in the implementation phase – the proposal will be modified, adapted and sometimes totally turned around. These characteristics of the political and administrative processes are not surprising, and from one point of view they should even be regarded as positive and something to be preserved (think of the democratic value of divided government and of checks and balances). Neither do they make it impossible to reach the desired result – for instance, the implementation of a user-charge system – but in order to do that a strategic way of thinking must be adopted by the proponents.

At a more theoretical level the same lesson can be drawn from Angela Liberatore's paper. As she points out, the arguments supporting the adoption of a new policy instrument can be any one of many different kinds (political, economic, legal, scientific or ethical), and, more importantly, can be supported more or less forcefully, by assumptions regarding human behaviour, institutions, technological development, nature and/or the context. The mix of arguments and assumptions used in any given context of decision making can be a factor explaining why or why not a desired decision is reached. When analyzing any specific process of policy formulation and adoption with this framework in mind one can reach surprising conclusions. For instance, in the case of the CO_2 tax, reconstructed by Angela Liberatore, the importance of setting a precedent of a European tax has played a not insignificant role in facilitating the acceptance of the same proposal at the EEC Commission level. Similar examples are by no means rare, and the lesson we can draw from them is that the designing of a strategy, including the analysis of the beliefs and the interest shared by the potential participants, is an important part of any policy proposal, including those that redefine the tool box of a policy actor.

If we now look at the content of policy instruments, i.e., the dimension(s) of the policy game they address, the first aspect we can try to modify is the number and the nature of the actors involved. As we have just mentioned the creation of the Environment Departments and/or Agencies in many countries has been regarded as a major policy instrument in the establishment of pollution control as an acceptable domain for governmental intervention. And, as we shall see in Peter Knoepfel's chapter, there are good reasons for thinking that, in the more or less near future, the trend towards developing special institutions for environmental protection will be reversed. The creation or the dissolution of national agencies, however, is not the only type of actor modification one can envisage.

As Nicola Pain suggests in here chapter, even in the time-honoured and well established field of jurisprudence, there are several possibilities which involve bringing in new actors which will have the effect of changing the way in which the environmental policy game is played. Two examples may be given. The first, and more traditional one, is the expansion of standing rights in courts in order to incorporate most forms of citizen action: increasing the number of subjects who can actually take action in defence of the environment multiplies enormously the ecological police function and decreases dramatically the level of certainty of the potential polluters have about their own legal rights (therefore opening up the field for a huge increase in collaborative and sometimes contractual solutions between the interested private parties). Of course, these effects are more likely to occur if governmental bodies or non-profit institutions provide financial resources or free legal aid to the citizen groups in order to make their standing rights both meaningful and not too expensive. The second example refers to the creation of specialist environmental courts, as has been done in New South Wales, Australia. One of the effects that such a development can produce is an increase in the scientific and technical knowledge available to the judicial system, making it possible, indeed likely, that the court decision will enter into more detail the merits of the case, for instance amending the administrative permit in order to prescribe a set of tougher conditions. In this sense the establishment of a special court, incorporating scientific and technical as well as legal knowledge can modify the rules of the game not only for the polluter, but also for an administrative agency which, under these circumstances, will not only know that its decisions can be challenged on procedural or otherwise legal grounds, but also that they will be reviewed on their merits and on the basis of their scientific and technical quality.

If the transformation of the set of actors involved is one of the possibilities open to the environmental policy reformers, the most usual species of policy instruments is, of course, the modification of the resources available to the different actors, in terms of legal rights, money, political consensus, information, know-how or time. The case presented by James Cameron and Will Wade-Gery is, from this point of view, rather unusual. The rise of the precautionary principle in international law-making is a good example of how it is possible to change the value of a given resource – in this case scientific knowledge about the harmful effects of some form of pollution – through a regulative intervention. The discourse of precaution boils down to the declaration that the uncertainty surrounding the effects of pollution, is not a good reason to avoid taking action. What the adoption of the precautionary principle in national or international legislation does is simply to say to the regulates that any existing scientific knowledge saying that there is no certainty of the potential harmful effect of a given substance or process is not grounds enough to given them the right to do whatever they want. The obligation to cease polluting (or, in the case of states, to regulate against polluters) stems from a sudden decrease in the value of the existing knowledge, equating uncertainty of risks with the probability of risks. Of course there is much more in the precautionary principle than this, and of course the very fact of reversing the burden of proof (the need to find the certainty of no-risk) changes dramatically the way in which the environmental game has been played until now, making the distinction between preventative and remedial instruments rather obsolete. The advantage of adopting a clear analytical approach, however, are such that one can design some new instruments using the analogy of the precautionary principle: for instance, in all the cases in which the issuing of a permit is conditional upon the provision of a certain amount of data, it seems possible to decrease the value of the existing information by simply requiring an additional series of measurements or information.

The last set of papers included in the collection refers mainly to the issue of the pattern of interactions. Here the hypothesis is that a change in the relationship between the actors involved in the policy game can bring about a desired effect. Joan Subirats' chapter, for instance, deals with the benefits that a wider adoption of policy evaluation can bring about. He argues that environmental policy provides a good example of where to try to use those forms of policy evaluation, such as pluralistic evaluation, which take into account the complexity of the real world, namely the fragmented reality of the policy community, the absence of clear rules of the game or shared problem definition, and the existence of conflict and disagreement over the goals and the causal theories. He suggests that the framework provided by the various techniques as well as the outcome of the evaluation exercise might (and probably will) modify not only the mutual attitude of the actors – and therefore the way in which they interact – but also the way in which they structure the problem and therefore the way in which they think and act. A widespread recourse to policy evaluation can trigger a learning process through which the different actors realize that the game they are involved in is a permanent one and that they have to learn to live with each other. In analytical terms policy evaluation is a policy instrument insofar as it is able to increase the information the different actors have about each other, and about the effects of the policy, therefore allowing a better definition of their strategies. Of course this doesn't mean that all conflict will be avoided, but it does have the advantage over more synoptic oriented approaches (like the one underlying the introduction of Environmental Impact Assessment) of accepting the basic fact of life that there are many possible, and politically and socially legitimate, ways of defining, structuring and solving environmental problems and that there may not be one best way to deal with term. Starting a process of mutual learning through the introduction of *ex ante* or *ex post* evaluation, seems worthwhile in creating a mutual understanding and less disruptive patterns of interaction (or rules of the game).

This is even more true in the case of the mediation processes discussed by Helmut Weidner. As he points out, the least one can say about the experience of the sort of authoritative instruments usually employed by governments in pollution control is that they cannot solve (and certainly have not solved) all the problems. A rethinking of the theory of the State, including a major cultural change on the part of the actors involved, is needed if environmental policy is to overcome the crisis it is experiencing in those countries, for instance Germany, which, having already developed a comprehensive regulatory system, are by now having difficulty making further progress towards preserving environment.

The point made by Helmut Weidner is theoretical and abstract in nature. If many cases of pollution control, and more generally of environmental conservation and improvement, are conflictual in nature, one must remember that the imposition of rules by the State is not the only, and in many cases not the best, solution, Most conflicts are solved not by someone imposing their own point of view, or by pushing the issue into court. A large number of environmental disputes can reach a satisfactory solution consensually, through the agreement of the interested parties, including the representatives of 'public interest'. Exactly as some agreements in the business world are reached through the intervention of a third party who is able to reduce the transaction costs of the process, in environmental policy one can also think, and by now several examples exists, of developing a mediation process through the appointment of a neutral arbitrator having the responsibility for the design of the appropriate procedure for dispute settlement and for presiding over its workings. It is obvious from what we have already said that the introduction of mediation in environmental matters is a policy instrument because it is directly aimed at redefining the patterns of interaction between the interested parties, creating the conditions of mutual trust that can facilitate the search for consensus. Even if the mediation procedures are usually aimed to find one-off solutions, their successful outcome can be interpreted as a learning process which is bound to have permanent effects on the way in which the actors will relate with each other in the future.

Not all interactions between the actors, however, are disputes between polluters, public agencies and those affected. A very important issue in the environmental policy making agenda is the need to find new ways of incorporating environmental policy into the different policy areas, both those closely linked to pollution control (health policy or land-use planning, for instance), and those potentially in conflict with it (agricultural, industrial, transport or energy policy, for instance). As Peter Knoepfel suggests in the final chapter of the book, this development will mean environmental policy is able to profit from the huge array of instruments these policies have developed over a long period of time. It is worthwhile emphasizing that the development of interpolicy cooperation is not an abstract process: the different organizational and legislative solutions must be discussed not only on their theoretical merits, but also in terms of their feasibility and of the outcomes they will provide, given a certain distribution of resources. For instance, even if it is a theoretically sound proposition that unifying the protection and use of natural resources will bring about better results – in terms of sustainability – than keeping the two functions apart, one has to ask oneself if, in a specific situation of time and space, in doing this one does not run the risk of subverting the protection function to the utilization function because, say, the forces behind the latter are stronger than the conservationist groups. The attempt to integrate ecological concerns into the different public policies, therefore, needs strategic thinking. But, one the other hand, if and when such a development will be possible there is little doubt that the general characteristics of the environmental policy game will change totally, and that the number of instruments available to the policy makers will increase dramatically.

3. Reforming Environmental Regulation: Some Ideas for a Research Agenda

The ideas put forward in the chapters of this book all support movement in the direction toward increasing the range of instruments available for environmental protection. There is no pretence whatsoever on the part of the authors of this book of having provided a full picture of what it is possible to do in this field in the foreseeable future. Not only have we consciously underplayed the promise apparent in the so-called 'economic' instruments (this theme is devoted another book in the same ESF program), but we have also omitted almost all references to the 'persuasive' or 'cultural' instruments that in the long run will probably prove decisive. In closing this introduction, however, it is worth stressing the beliefs we share as far as the general question of environmental policy is concerned. None of these beliefs is very original, but sometimes the debate in academic as well as governmental circles tends to underestimate the relevance of some simple truths.

The first, as we have already pointed out, is the opportunity, indeed the need, to integrate the discussion about environmental policy design into the more general framework provided by the analytical traditions of international and constitutional law, public administration and public policy analysis. We have to fight against the temptation to build up environmental policy as a totally separate field of research, thus losing the richness of ideas accumulated in the study of other policy fields and other institutional arrangements, as well as making more difficult the accumulation of knowledge due to the lack of an analytical approach to the solution of collective problems.

The second belief we share is the fact that comparing different instruments, or families of instruments, in order to affirm the superiority of one or another is a less promising research strategy than searching for possible combinations of the different policy instruments or the careful investigation of the conditions under which each of them works best (and this includes, obviously, both the decisional costs in adopting, and the enforcement costs in getting compliance from the polluters). Environmental policy instruments are not an end in themselves, but a means of achieving better protection.

A third point worth stressing has already been referred to in the first paragraph when addressing the question of the so-called globalization of environmental policy and of the challenges this poses to the current standard operating procedures. We firmly believe that the future research agenda, at least as far as the economically advanced countries are concerned, should put a strong priority on the question of the 'ecologization of the State'. This implies the question of institution building, including the opportunity for new constitutional norms, different administrative and technical structures, a better balance between state and market regulation, a new equilibrium between national and regional and local interventions, etc. We will come back again to some of these issue, but what we want to emphasize here is the need to adopt a comprehensive rather than issue-specific approach. It is really not enough to think in terms of a second or third generation of environmental policy. One should actually think in terms of the need to incorporate environmental concerns into the different policy fields: the problem is less the new environmental policy and more the new agricultural, industrial, transport policies. It is possible that the inclusion of the environmental concerns will in some cases help to solve problems already inherent in these policies (think of the Common Agricultural Policy, for instance), and it is likely that in other cases the reverse will occur. But this is not a matter of choice: exactly as the national (and soon supranational) institutions are and will be responsible for the economic and social development of their countries, they will also be held responsible for keeping a correct balance of the existing natural resources. The possible alternative, even in the short run, is a progressive marginalization of environmental policy because of the constraints it poses on individual and economic freedom.

This brings us to a different, and much more ambitious, line of reasoning. Is it possible to continue developing indefinitely a policy which is essentially negative in character, putting more and more stringent limitations on human behaviour, and whose only goal is to preserve the existing situation? In other words, is there the need to think in more positive, one could say active, terms as far as the question of the environment is concerned? This is of course more a question of what the legitimate values to pursue are to be, than one of what the best instruments in fact are. There is little doubt, however, that if we had the technological ability to modify the environment in order to get a 'better' one (for instance in order to avoid droughts or to prevent volcanic eruptions) and if this could be considered a legitimate political goal, then we would have the problem of finding the most appropriate policy instruments. The discussion of the implications of such a revision is by no means a luxury exercise. On one hand one should think of the far reaching consequences that a 'positive vision' of the environment could have on the actor's behaviour. One the other hand it is possible that we will have to face some dilemma of this sort in the short run: think for instance of the possible relationship between the development of bio-engineering and the preservation of bio-diversity. How to regulate such a matter? To add this point to the research agenda seems to us an urgent matter.

If this desire for a further inquiry into the complex questions of the ecologization of the state and into the need for a positive environmental vision seems to emphasize too much the general and abstract aspects of environmental policy (but I hope to have made it clear that in our opinion these are indeed very practical and urgent issues), the last point we want to make is, on the contrary, much more down to earth, at least at a first look.

Our very deeply shared belief, stemming from our professional knowledge of environmental policy and of governmental activity in general, is that there are important limits to the effectiveness of public regulation in the field of environment. One does not need to share the ideological underpinnings of the deregulation debate in some countries in order to understand that there are many risks behind a further increase in the use of command and control type of tools. The usual argument used in the reform of regulation debate stresses the hidden costs of regulation and the distributive distortions it can entail, as well as the perverse effect over economic competition and the free market it can have. Of course an answer to this is the one implied in the sustainability principle, i.e., the observation that the alternative is to consider natural resources as not being totally free, because such a view means allocating the costs to the future generations. A good solution could be to develop techniques able to reveal the economic impacts of regulatory instruments, as has already been done in several countries. From this point of view the evolution of regulatory accounting and budgeting is another point to be added to the research as well as the policy agenda.

However, when we are worried about the effects of a big increase of environmental public regulation, we think mainly of one different aspect that in a certain measure applies also to the so-called economic instruments, namely the occurrence of 'regulatory inflation', or, to use a different metaphor, of the 'entropy' in the regulatory system. By these two metaphors we want to describe a situation in which the production of new norms decreases the value of the existing ones, because of the sheer complexity of the overall policy. This implies that, while the decisional costs of a new, possibly very specific, regulation are becoming increasingly small, the enforcement costs become larger and larger, not least because the regulates do not know precisely what is expected of them, and because the public agencies are structurally too understaffed to allow real policy implementation to occur.

These dangers are not theoretical. The effects of the past emphasis on deregulation can be seen mainly in business regulations, while the body of laws and norms aimed at protecting the environment has become larger and larger. There are several reasons for this continuous increase. Quite apart from the 'good reasons' (including, for instance, a wider application of the precautionary principle) there are also some perverse effects of the post-welfare policies adopted in many countries.

Privatization is a case in point. The fact of giving back to the market some or all the economic activities previously in the hands of the state calls for more stringent regulations, and if the activity is related to the use of natural resources it calls for more environmental regulation. The obvious example here is that of water management, but many other examples can be pointed out. A more subtle and rather paradoxical process going in the same direction is the reduction of public spending, which reduces the range of policy instruments available to the existing public bureaucracies. They therefore find the recourse to regulation a viable way of maintaining their power in the new circumstances.

All this boils down to the need to find alternatives to the existing tradition, also within the regulation itself. This point needs to be put very firmly on the research agenda. In my opinion the investigation should go in two different but complementary directions. The first concerns the genesis of regulation, i.e., the problem of the regulator. It is by no means obvious that the one and only way of establishing binding obligations is through authoritarian imposition by an external actor, the state or the government. International law is an obvious example of a system of rules evolving through collective agreements, and this has proved quite effective in providing stringent guidelines to the regulates (think of the precautionary principle). Why the same is not true as far as domestic environmental regulation is concerned? One of the best known reasons is the idea incorporated in models like the 'tragedy of the commons', the 'prisoner's dilemma' or Mancur Olson's "logic of collective action". This line of reasoning stresses the difficulty, indeed the logical impossibility, of avoiding free riding and non-cooperative behaviour. Of course this is a very powerful objection to the development of self-regulation, but one that has the disadvantage of not taking into account many facts of life, as, for instance, the fact that in some common pool resources there is indeed self-regulation and it works quite well (for some examples see OSTROM 1990, which provides a stimulating and balanced approach to the question of regulating common pool resources). This can be because of the lack of suitable alternatives, for instance because the decisional as well as the enforcement costs involved in public regulation are simply too high and cannot be afforded. Sometimes this can lead to the real tragedy, and sometimes it can stimulate a cooperative approach.

In the case of environmental protection, one the other hand, the growing public awareness of the importance of the problem has probably changed both the ethical attitude and the self-interest perception of the people. The dangers of a non-disciplined use of the natural resources are by now quite well known, and, as I suggested in the first paragraph, the environmentalist ideology, at least in this part of the century, has become common sense.

What the above mentioned theoretical models do imply is that contemporary citizens are not the myopic fools, narrowly egoistic and totally ignorant of the consequences of their actions, presupposed by other models. Nor do they exist in a cell like the participants in the prisoner's dilemma, but can communicate and discuss between themselves in order to reach the most satisfactory solution to their common problems. This doesn't mean, and no one here suggests it, that self-regulation is the one best way to optimize environmental protection or that governmental intervention, by means of command and control or any other kind of instruments, is the only solution. What we do point to is the need for a careful, wide and interdisciplinary research program on the promise of self-regulation, especially aimed at investigating the circumstances under which, and the processes through which, it can emerge, and where and when it works best in securing the preservation of the natural resources. For instance there is little doubt that the sheer spatial and social dimension of the environmental problem plays a role here, facilitating or hindering the emergence of self-regulation. While for global problems there seems to be no alternative but to look for cooperation, the same is not true as far as domestic policy is concerned. But again one could argue that to fragment the problems, wherever and whenever possible, in order to enhance self-regulation is not a bad idea at all. Take for instance the problem of waste disposal, indeed of waste reduction; it is more than possible that, once there is a common understanding, within a given community, of the seriousness of the problem, once it is impossible to externalize the negative effects, and once it is clear that to find a solution is not someone else's business, then a satisfactory solution can emerge, which will possibly be enforced at a financial and social cost that is but a fraction of what will be needed in case of state intervention.

The need to investigate the potential for self-regulation includes also the need to draw boundaries – someone would say constitutional limits – to the possibility of government intervention. Here something like the subsidiarily principle should be developed in order to provide for a dynamic adaptation of the sphere of influence, indeed of the legal competences, of potentially conflicting sources of binding obligations, the international community and the individual sovereign states, the federations and their member states, the central administration and the local and regional governments, the regulatory state and the self-regulating society. This leads us back to the question of the ecologization of the state, and we want to suggest here that one way of achieving better policy coordination, indeed the inclusion of environmental concerns into the various policies, is a process of decentralization, through the spatial segmentation of the environmental problems, when of course this is actually possible.

The problem of the genesis of regulation is not the only point to be addressed in the future research agenda. One should also tackle the difficult issue of the regulatory content, that is of the flexibility of regulations. The problem of flexibility is at the basis of the whole discussion about the relative advantages of economic and financial instruments over command and control type of tools. The advocates of the economization of environmental policy observe that uniform standards do not take into account the different costs of depolluting which the different industries have to bear. The same level of environmental protection can be reached in a more efficient way – they argue – if we leave the individual industries free to decide whether, given their cost structure, they prefer to maintain the same level of emissions, paying accordingly, or to depollute. This line of reasoning implies in the first place that the goal of environmental policy is to keep the level of pollution under a given threshold and in the second place that command and control instruments are by their nature rigid, i.e., uniform in all the individual cases. The former assumption certainly is at least partly realistic in describing the first generation of environmental policies. One could say that the philosophy of stringent and uniform emission and immission standards was toxicologic in nature. Pollution, it would suggest, is a poison, but if the dose is small enough there is no danger. Every emission above the fixed threshold is bad, and should be considered illegal. Every emission below the standard is would not be considered a toxic emission from the public policy point of view. Is not surprising, therefore, that all the public discussion was about the level of the regulatory standards, running the complementary risks of setting them too low, threatening the environment, or too high, unreasonably penalizing the economy.

Quite apart from the question of setting standing, however, one has to ask oneself if the "toxicologic approach" is capable of providing a foundation for the policy. The concept of sustainable development, on one hand, and even more the precautionary principle, on the other hand, seem to suggest that this is not the case. They imply that one has to try to limit environmental impacts as far as possible, even without any scientific certainty about the adverse effects of the behaviours. In a sense, in this new approach there are no thresholds, or, more precisely, all the behaviours are at the same time well below the thresholds themselves, because the progress in depollution has already ruled out all the emissions whose danger is scientifically certain or probable, and well above the threshold, because precaution aims at curbing every emission, even if there is no firm scientific evidence about the non-absorbability by the environment. It is obvious that in this situation a totally new generation of regulatory policy instruments are to be developed (and probably also new economic instruments).

In fact the second assumption referred to above, that command and control instruments are rigid in nature, i.e. uniform in all the individual cases, is not correct. In the real world this apparent inflexibility, and even the seeming uniformity, has been tempered in many ways. Think, for instance, of the principle of the Best Practicable Means, or of the Best Available Technology. They have been in use a long time, and have permitted a dynamic adaptation of the rules to the needs of technological development, as well as an individualization of the regulations according to the different circumstances of time and space. Of course increasing the discretion of the regulator entails the possibility of mistake, but on one hand the enforcing agency usually is a technical bureaucracy, less prone to error than, say, a parliament, and on the other hand being too strict or too lax in regulating individual behaviour is less dangerous than imposing an inappropriate general standard or creating a bad tax structure. Regardless of what non-lawyers think, there has always been room within the regulatory system for passing out differential treatment to each of the various target groups, and also to individuals. Of course this has advantages as well as disadvantages, and one of the things we have to study and to learn is how to profit from the former and how to avoid the latter.

Increasing the level of technical and bureaucratic discretion given to the agencies is not however the only way of creating more flexible regulations. Another, much more original, means is by adhering to the notion of "averaging". The example here is a regulation employed by the American car industry, where, as far as the pollution characteristics of the cars are concerned, the manufacturers can average emissions across their fleets in order to meet the standard. This means that they can put on the market a model whose emissions are above the standard, if and only if they can produce and sell enough cars having emissions below the standard such that the average emission of all cars produced meets or is below the standard as well. In the case of California, where legislation which will effect a sharp reduction of emissions is scheduled between 1994 and 2003, they can also carry over their pollution credits from one year to the next, subject to a "devaluation" rate of 25% (Sperling, 1991).

Starting from these examples one could try to develop a system of industry self-regulation. Every individual firm has or can have many different factories, each of which can be the source of one or more sorts of pollution (air, water, soil, noise, waste, etc.). Assuming (1) that there is no breaching of any of the "absolute standards" (i.e., that on no one aspect the firm is behaving in a way bound to seriously endanger the environment), and (2) that there is a system for measuring and comparing all forms of environment impact, a system of environmental accounting which can be done at the micro-level, how then might we go about obtaining further reductions of the total impact, bearing in mind guidelines provided by the precautionary principle? One realistic method would be for the public authorities (but also a self-regulating industrial association) to define a path of overall reduction of environmental impact, for instance prescribing a 2% annual decrease in total consumption of natural resources, and at the same time requiring the creation of an environmental audit system which would provide a minimum set of measurements and information, and a review of the audit by some independent private body. It will be left to the individual firm to decide where to make their changes *vis-à-vis* the environmental impact of their various activities: one could think of alternative measures as, for instance, a change in the packaging in order to reduce the waste problem, a shift to different ways to deliver the goods to the customers, new production processes, etc.

The key factor, here, is the development of environmental audit systems and, more generally, of environmental accounting. In my personal opinion the promise of such instruments in bringing about a new generation of policies, based on flexible regulation, is potentially enormous. The first timid introduction of such tools at the EEC level (the regulation on the eco-label and the proposal for creating environmental auditors) is an important step in the right direction, even if these first instances seem a bit cumbersome and still very "regulatory" in the old tradition. Of course these instruments cannot, and will not, solve all the problems – but no instrument can. Their potential is high in all the cases in which the level of environmental consciousness and the technical capability of the target group is strong enough, and this applies to most industrial sources in many European countries.

There are indeed powerful arguments in favour of a "global impact approach". The most obvious is that, in the real world, there is neither the need, nor the possibility, to impose indefinitely more and more stringent standards to address the different forms of pollution, and the same is true as far as the economic instruments are concerned. The attempt to create, within a single area of environmental concern, a sort of "internal market", in which all forms of pollution can be accounted for, and the costs of abatement compared, seems, below a certain threshold, a much more viable solution than traditional regulation, or than trying to create markets for pollution permits between different polluters, experiments that until now have been rather disappointing in their results, i.e. in getting an actual trade.

Here again we are back to the issue of which are the main factors motivating polluters' behaviour (short-term profit maximization, share of the market, or technical progress). Unless an inquiry of this sort is incorporated into environmental policy design, i.e. unless a new generation of policy instruments based on a realistic assessment of the new and old actors of the game, of their resources as well as of their patterns of interaction, is developed, any further progress of the public endeavours in bringing about a better protection, meeting the challenges of sustainability and globalization, will be very difficult indeed.

References

Dunn W.M. (1981) *Public Policy Analysis – An Introduction*, Prentice Hall, Englewood Cliffs.

Haas P.M. (1990) *Saving the Mediterranean – The Politics of International Environmental Cooperation*, Columbia University Press, New York.

Mazmanian D. (1991) California green – Toward the new energy paradigm, paper presented at the international conference "Implementing Environmental Policies by Means Of Interpolicy Cooperation", Crans Montana, September, mimeo.

Ostrom E. (1990) *Governing the Commons – The Evolution of Institutions for Collective Action*, Cambridge University Press, Cambridge.

Sperling D. (1991) *Marketable Credits for Vehicle Emissions in California*, paper presented at the international conference "Implementing Environmental Policies By Means Of Interpolicy Cooperation", Cans Montana, September, mimeo.

Tummers L.J.G. (1991) Industry's View on the Use of Voluntary Commitments as an Alternative to Regulations to Comply With Environmental Standards, in M.Dubini (ed), *Environmental Regulation: Approaches for the 1990s and Beyond*, Istituto per l'Ambiente, Milano.

2. A Classification of Instruments for Environmental Policy

P.-J. KLOK
University of Twente, Enshede, The Netherlands

1. Inroduction

The purpose of this chapter is to develop a classification of the instruments for environmental policy. Of course the intention to classify instruments is not new and many attempts have been made already. In our view most of these previous attempts suffer from one or more of the following shortcomings:

- The instruments are classified according to one or a few differentiating characteristics, neglecting characteristics that the instruments have in common. As an example one can think of the omnipresent distinction made between 'regulative' and 'financial' or 'economic' instruments, that neglects the fact that almost all regulative instruments have financial or comparable elements (sanctions) and the fact that almost all financial instruments are formulated in terms of legal rules (Klok, 1991);
- The classifications concentrate on elaborate and more or less systematic schemes for more traditional instruments (like regulative and economic instruments), without developing a comparably systematic scheme for instruments that have come into discussion more recently (like different forms of communication, consensus building and institutional (re)arrangements);
- The classifications are more or less closed systems, that have considerable trouble incorporating new instruments that arise from the occasional shows of creativity on the part of political or scientific actors.

In order to overcome these shortcomings the following conditions for an adequate classification should be taken into account:

- The classification should concentrate on both the differences and the similarities of policy instruments. This means that the classification is not a set of sharply delimitated and exclusive cells, but a method for the characterisation of policy instruments (showing differences and similarities);

Bruno Dente (ed), Environmental Policy in Search of New Instruments, 21–36.

- The classification should be broad and systematic;
- The classification should be an open system, allowing newly developed instruments to be characterised as well as existing ones.

Before a classification is developed, a major distinction should be made in advance. On a general level of abstraction it can be said that environmental problems are the result of the interaction of human activity and environmental processes. This means that policy instruments can be directed towards changing human activity (behaviour) or towards changing environmental processes. If a policy actor decides to intervene directly into environmental processes, for instance by cleaning contaminated soil, treating waste water or dumping lime into acid lakes, he uses fundamentally different instruments from the ones that he would use to influence human activity. This chapter will concentrate on policy instruments that are aimed at influencing human activity.

2. A Simple Model of Human Activity

If one wants to classify the ways in which a policy actor tries to influence human activity, it might be fruitful to start with a model of human activity. This step seems even more necessary if we realise that the attempt to influence human activity is a form of human activity itself. When we speak of *human* activity, we mean activity by either a single individual or by an organisation. Form now on the word *actor* will be used, which can mean either an individual or an organisation.

At the most general level of abstraction one could say that almost all forms of activity (behaviour) need at least some sort of *resource* and some form of *motivation*. Even the simplest forms of biological activity need resources like food and water. While there are some forms of activity that seem 'internally motivated', like breathing and the beating of a heart, most other human activity would not be likely to take place without some form of motivation.

If one looks more closely at the different forms of *resources* that an actor might have, it is possible to start with the classic (economic) distinction between *physical goods* and *skilled people*.[1] Physical goods might range from buildings, complex machinery and simple tools, towards food and other 'basic' resources like energy, water and clean air. Skilled people as a resource for organisations refers both to the presence of people and to their skills and experience. In the case of an individual this resource refers to his or her skills and experience.

A third form of resources might be called *information*. It is one of the outstanding capacities of man that he can conceptualise his experience in language, store this information either in his brain or in signs and use it when he thinks it is right to do so. In this sense information plays a key role in most individual and organisational activity.

A fourth essential resource for any activity is *time*. Almost by definition there is a relation between time and activity, in the sense that without time it would be impossible to conceptualise an activity as some sort of changing stage of the world. In western society there seems to be a continuous speeding up of activity and time is likely to become an increasingly scarce resource.[2] When speaking on the amount of time an actor has at his disposal, it might be useful to speak of his *time-budget*.

The four types of resource mentioned above could be enough for the activity of a single actor. A carpenter who wants to make a chair might use raw material and tools (physical goods), his skills, information (for instance on the type of material he needs) and time in order to make that chair. In practice however, most actors live in a social environment filled with other actors. Most activities of actors are related to activities of others. Due to the fact that most 'action' is part of some form of 'interaction' it is necessary to distinguish some additional types of resource. The most obvious 'social' resource is *money*. The conceptualisation of money as a resource for human action is so obvious, that it needs no examples.

As much as the interaction of actors needs a facilitator ('money makes the world go round'), it also needs a restrictor. *Legal rules* generally perform the function of allowing some forms of activity to an actor (and thereby preventing others from intervening in this activity) and restricting other forms of activity for an actor (and allowing selected other actors to intervene if the activity is performed despite the prohibitive rule). In the sense that legal rules allow an actor to choose particular activities and do not allow him to choose other activities, these rules can be seen as an important resource for human activity. If one speaks of resources that enable an actor to perform an activity it would be better to refer to these resources as *legal rights*.

The last type of social resource 'par excellence' that is distinguished could be labelled *consensus, authority* or *trust*.[3] Just as the very concept of money depends on consensus or trust, human interaction depends on them as well. If the only way to produce predictable interaction was to actually apply legal rules, then the social system would collapse. If an actor wants to sell a product to another actor (which can be seen as an activity), then this activity is highly dependent on trust. Without some degree of trust that the product is valuable (which depends on consensus on what can be seen as a 'good' or 'valuable' product), the sale will not take place. In this sense trust functions both as a restrictor and as a facilitator. The need for trust on the part of the seller functions as a restrictor, for the selling of bad products would reduce his trust and make this activity in the future impossible. For the buyer it functions as a lubricant, for his trust in the product discharges him of the resources consuming activity of ensuring good products by applying legal rules.

After the description of different types of resources it is necessary to pay some attention to *motivation* as the other element in our model of human activity. A general distinction that could be made is the one between *internal goals* and *external goals*. Perhaps the most obvious internal goal is 'survival' and the closely related ability to perform activities. Since we have seen that the ability to perform activities is highly dependent on the actors' resources, it follows that a prime motive for action is the possession of a certain amount of resources. While some might believe that this is the only human and organizational motive for activity, we believe that there are other internal human 'needs' that motivate action and we assume that actors can be motivated by external goals. In the case of governmental organizations and most certainly in the case of actors ensuring environmental policy, the motivation for action seems to be an external goal (for instance 'environmental quality'). Because the concept of motivation is not as central to our classification as the concept of resources, the discussion on motivation is dropped here, and a full classification of different types of human motives is not developed.

After this short sidestep on motivation it is useful to dig somewhat deeper into the concept of resources. Some reference has already been made to the fact that resources are in most cases scarce, which means an actor has a limited amount of these resources. It was also stated that the amount of resources may vary. This brings us to the central concepts of 'stock' and 'flow' of resources. *Stock* refers to the amount of resources an actor has at his disposal. Stock can be at different levels for each resource. *Flow* refers to variation in the amount of resources. When thinking of different types of flow, it is obvious to make an initial distinction between *positive* flow (growing stock) and *negative* flow (declining stock). In a model where an actor is supposed to live with other actors in an environment, it is possible to distinguish at least six fundamental types of flow:

- the intake of resources from other actors;
- the intake of resources from the environment;
- the production of resources by the actor;
- the use of resources by the actor;
- the giving out of resources to other actors;
- the giving out of resources to the environment.

From the perspective of the actor the first three types of flow are positive (growing stock) whereas the last three are negative (declining stock). Since these forms of flow are changes in stock and these changes are generally not 'automatic', each type of flow normally refers to an activity of the actor. Resources don't usually 'fly' towards the actor, but he has to bring them in, which is an activity. In economic life the activities of taking resources in from others and giving them out to others are usually combined. This combination is called 'trade'. This does not mean that these activities are always combined. Sometimes there is 'taking without giving' (for instance 'theft') and sometimes there is 'giving without taking' (for instance 'charity'). The production and use of resources is in economic theory generally divided between 'production' and 'consumption'. In the model developed here, there won't be such a sharp distinction, for any activity, even if it is aimed at producing resources, needs and uses resources. In this sense the result of production is the amount of the different resources generated minus the amount of resources used in producing them.

Given the fact that one looks at policy instruments for environmental policy, it seems obvious to have a special interest in the relation between the resources of an actor and the *environment*. The taking in of resources from the environment refers to what could be called 'basic' or 'natural' resources (physical goods), like minerals, fossil fuel, water and air. These natural resources taken from the environment are to some extent used in production, consumption and trade. They eventually return to the environment in the form of waste *emissions* (solid, fluid or aeriform waste). To a certain extent this waste can be regenerated into basic resources by the environment itself, without intervention (e.g. CO_2 into oxygen). In this way it is possible to introduce the concept of 'sustainability' of resources. Sustainability of basic resources is reached when human activity takes no more resources than the environment can generate and emits no more waste than the environment can bear or can regenerate into resources.

The model of human activity is represented schematically in Figure 1. It shows

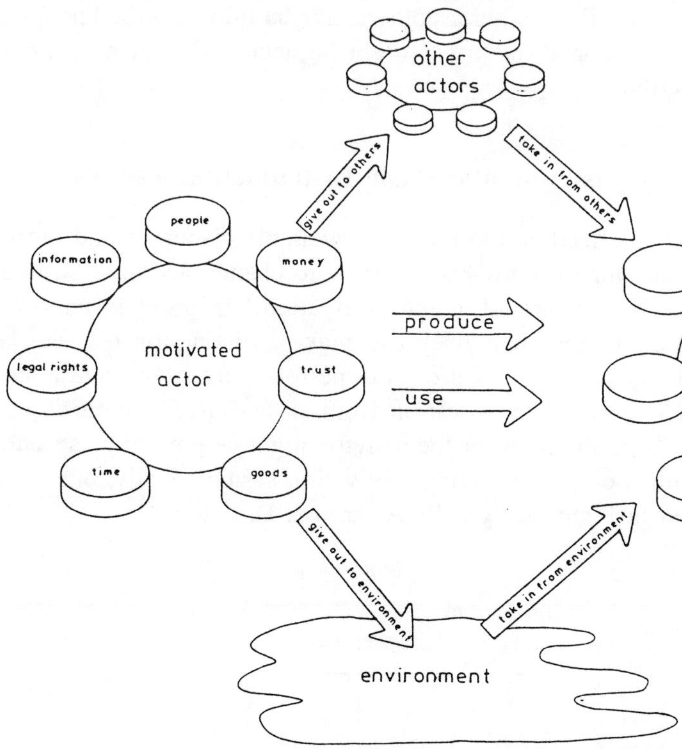

Fig. 1: A simple model of human activity.

a motivated actor with his seven resources, represented by cylinders, giving some indication of stock. In between the two representations of the actor are the six activities that refer to the six forms of flow in resources. The second representation of the actor shows him with his new stock of resources and his new level of motivation (this level is not represented in a graphic way).

Before turning to policy instruments, some additional remarks can be made on the relation between resources and activity. First, it is generally the case that a certain *combination of resources* is needed for an activity. An activity will always take time and people, and will mostly depend on some amount of goods, money, information, legal rights and consensus/trust. To a certain level however the amount of the different resources that is used may vary. There is a degree of *substitutability* of the resources. For instance the same activity could be done in little time with many people and little money, in little time with few people and much money, or in much time with few people and little money. The different and alternative combinations of resources that enable an actor to perform a certain activity would therefore be a major focus of analysis. A second important remark concerns the question whether resources that are necessary for an activity actually 'get used' in this activity (decline in stock). Some resources, like materials (physical goods), time and money usually get used (decline), whereas others, like buildings, skilled people, information, legal rights and consensus/trust might be necessary, but might not decline in stock by the activity.

3. Resources and Policy Instruments: Basic Instrumental Tactics

If human activity is conceptualized as the result of available resources and motivation, then the obvious answer to the question 'How to change activity?' would be: 'Change the available resources and/or the motivation.' In order to analyze the ways one can influence activity it is useful to make a distinction between a situation where one wants an actor (target group) to perform an activity and a situation where one would like an actor to refrain from an activity. In both cases, the resources needed and the motivation for the activity might be present or absent. Combining these possibilities of resources and motivation systematically, one gets four situations represented in Figure 2 by cells A through D.

	Motivation	
Resources	Present	Absent
Present	A	B
Absent	C	D

Fig. 2: Resources and motivation related to the performance of activity.

If the policy aims at performing a certain activity, then this activity will generally be performed by the actor if both the needed resources and motivation are present (cell A). If this is already the case, no attempt to influence activity might be necessary. In the other cases (cells B, C and D), the question is how to reach cell A. This shift might be reached most easily in a situation where motivation is present but the needed resources are absent (cell C). In this case it would be enough to provide the actor with the lacking resources to get to cell A. *Provision of resources* could therefore be distinguished as a first instrumental tactic.

Getting from cell B to cell A would require a change in motivation. The first way to influence motivation might be to try to convince the actor that he should change his motives. This would however be a special case of provision of resources (information) and consequently does not provide a new form of instrumental tactic. Given the fact that resources are scarce and necessary for activity and thus valuable for the actor (and thus also an important motivation), an obvious and often used way to change motivation is to alter his stock of resources depending on his activity. The first instrumental tactic that uses this mechanism could be called *conditional provision of resources*. If a policy promises to give the actor in the target group resources when he performs the activity, the value of the promised resources functions as a motivation to perform the activity. In a comparable way the policy actor can threaten the target group with taking resources from them if they don't perform the activity. This time the value of the resources the target group might lose will function as a motivation for the activity. This tactic can be called *conditional deprivation of resources*.

Getting from cell D to cell A would require a change in both resources and motivation. This does not lead to new forms of instrumental tactics, for one could use either conditional provision of resources (which could perform both functions at the same time) or a combination of the other tactics.

From the analysis of how to get an actor to perform an activity we now turn to the situation where one wants the actor to refrain from a certain activity. Like in the first part of the analysis, the only situation where the (in this circumstance unwanted) activity can be expected is the situation where both the needed resources and the motivation are present (cell A). In this situation the instrumental tactics are used to reach *any* of the other cells (B, C or D). In all three cases the actor will refrain from the activity. The first possible way to get the desired result would be to change to cell C, where the actor lacks sufficient amount of resources to perform the activity. Taking away some of the resources needed by the actor would do the job. This tactic could be called *deprivation of resources*. The movement towards the other cells (B or D) does not provide any new instrumental tactics. Conditional provision (if you refrain from . . . you will get . . .) or deprivation (if you do . . . , I will take . . .) of resources can change motivation in a way that is comparable to the ways already mentioned.

At this moment it might be useful to give some examples of the four instrumental tactics mentioned above. Before doing so it is important to stress that these tactics are not the only ones possible. In terms of the already distinguished six types of flow of resources, the four instrumental tactics can be seen as the ones that are the result of the 'taking' and 'giving' of resources between the policy actor and the members of the target group (those whose activities the policy actor wants to change). After the examples, some instrumental tactics will be distinguished that refer to the exchange of resources ('give' and 'take') between the target group and other actors than the policy actor and other tactics of a more 'indirect nature'.

When thinking of examples of *provision of resources* one must keep in mind that this instrumental tactic is applicable in situations where the motivation for the activity already exists or is the result of other tactics. If one imagines an environmentalist group that would like to take up the maintenance of a certain forest, then a policy actor might provide this group with machinery (physical goods), a subsidy (money), experts from the policy actor (skilled people), advice (information) and/or the legal right to do maintenance in the forest. Of course many different combinations of these resources are possible. One might have more trouble conceptualizing the giving of 'time' and 'consensus/trust' in this situation. Time itself could be conceptualized as a continuous and invariable flow. The amount of time that is needed for an activity is however strongly dependent on the amount of other resources that an actor has at his disposal. In this sense for instance the provision of goods, money, people and information might greatly reduce the time that the actor needs to perform the activity. In this way the actor has (comparably) more time at his disposal. The time that an actor has at his disposal for an activity can also be conceptualized in terms of the moment that a certain activity has to be finished. In this sense time can be 'given' to an actor by postponing the moment that an activity has to be finished. The resource 'consensus/trust' has by definition an interpersonal and conceptual (non-physical) character. This means that this resource is not exchanged in measurable quantities, but is the result of shared activities. Consensus/trust might increase because of interactions that provide valuable results for the actor and might decline because of interactions that have negatively valued results.

The second instrumental tactic was called *conditional provision of resources*. This time the provision depends explicitly on the activity that is wanted from the actor (by the policy actor). The provision of goods and money are often based on the stated intention of an activity (for instance in scientific research projects) or on the demonstration that an actor has already performed an activity (for instance subsidies on investment). In most cases the instruments also contain legal rules that state explicitly in which cases the conditions for the provision are met. All other resources might be given on a conditional basis. An often used means of conditional provision of legal rights is found in many license or permit systems. The right to perform an activity (the licence/permit) is given only if certain conditions are met (for instance measures are taken to prevent damage to the environment).

The *conditional deprivation of resources* might perform the same motivational function as conditional provision. Conditional deprivation of goods, money or people are often seen in sanctions that accompany legal rules. If an actor does not comply with the rules he might loose his car (in traffic regulations), he might have to pay a fine or he might loose his manager (in case of imprisonment of the responsible officers in 'organizational crime'). Of course 'sanctions' are not the only possible way of conditional deprivation of these resources. Conditional taxation is another example. Legal rights and time can also be taken in a conditional way (if you don't do this, we will forbid you legally; or if you don't do that we will give you less time to do . . .). Conditional taking of consensus/trust can be done by threatening an actor with publicing information about his unreliability (if you don't do this, we will inform the press about your environmentally unfriendly behaviour). The reduction of information is somewhat more difficult to conceptualize. Thanks to the ease of copying information (both from the brain and from physical forms of storage) it is relatively hard to actually reduce stock of information. In both cases it is possible to think of examples (Like 'brainwashing' and destruction of archives), but it is not very likely that these 'instrumental tactics' will be used on a large scale.

Just as resources can be taken on certain conditions, there can also be examples of unconditional *deprivation of resources*. In some cases (like imprisonment and the taking of a car), the actual execution of a sanction might not have a deterrent effect (as is the case in conditional deprivation), but might have the function of actually making a certain activity impossible. Of course the question in this case is whether the deprivation of resources is sufficient to deprive the actor of all possible combinations of resources that might enable him to perform the activity. The 'classic' example of this case may be the deprivation of the legal right to perform a certain activity (prohibition of an activity). This deprivation changes the resource 'legal rights', but this does not mean that the activity is really impossible for the actor, as can be seen by day in day out violation of rules.

In the section above some examples are given of the instrumental tactics that were described before. Some additional remarks can be made.

First, it follows from the section on flow of resources that all flow generally needs activity. If we conceptualize policy instruments in terms of (conditional) flow of resources, then it follows that policy instruments also need activity. Activity from the policy actor is needed to actually realise flow of resources. We will refer to this activity as *implementation*.

Second, in order to make use of policy instruments the policy actor himself *needs resources*. The policy actor needs resources, because the use of policy instruments needs activity (implementation) and activity needs resources. Beside that, the policy actor needs resources as far as some instruments consist, at least to some extent, of giving resources to the target group. The distinction for a policy actor between getting resources in (in order to function as an actor) and giving resources out (in order to reach policy goals) is somewhat akin to the distinction between 'detectors' and 'effectors' presented by Hood (Hood, 1983). From the four distinguished instrumental tactics one can however see that some of these tactics may fulfil both functions at the same time ((conditional) taking of resources). From the point of view of the policy actor however, policy instruments can be characterised as fulfilling either one or both of these functions.

Third, the difference between (conditional) provision and (conditional) deprivation of resources seems to account to a great extent for the *valuation of the instrument by the target group*. It seems obvious that the provision of (valued) resources would be more welcome to the target group than the deprivation of resources. This difference refers more or less to the distinction made by others between 'enlarging' and 'restricting' instruments (Van der Doelen, 1989).

Finally, some attention should be given to the fact that the 'policy actor' is usually a governmental organization. This results in some cases in a special position concerning the flow of certain resources. Due to the legal system in a country the policy actor can have the ability to change the legal rights of the target group. The competences for legal rule-making will however generally be restricted to certain activities of the target group. Where they are present for some forms of behavior of the target group, they might be absent for other form of behavior. The ability of the governmental policy actor to use money as an instrumental tactic is severely restricted by his budgetary constraints. One the other hand he might be able to raise money through his tax-raising competences. In short, when analyzing the policy instruments available to a policy actor, it is necessary to give attention to his abilities and constraints concerning the flow of resources related to the target group.

4. Multi-layer Instrumental Tactics

In the section above we have discussed four basic instrumental tactics. These tactics can be used to influence any type of activity. However, not all activities of an actor are relevant to environmental problems or protection. The most straightforward way to influence environmental problems that are related to the activity of an actor, is to influence activities that have direct environmental consequences. In general, environmental problems will be related to the flow of physical goods. Each type of flow of physical goods can have direct environmental consequences. In this sense a policy instrument can also be characterized by the type of flow of physical goods (activity) that is its focus. In this way we have instruments addressed at taking in goods from other actors or from environment, producing goods, using goods, giving goods (products of waste) to others and giving goods to the environment (emissions). All basic instrumental tactics can be addressed, in different combinations of resources involved, to these six forms of flow of physical goods.

The basic instrumental tactics are however not the only possible way to influence the resources and motivation that are relevant to the six flows of physical goods. The exchange of resources with the policy actor accounts generally for only a minor part of the flow of resources to, from and within an actor. A first alternative instrumental tactic would be to influence activities of the actor that are not directly relevant to the environment, but are relevant to the resources and/or motivation that are related to environmentally relevant activities. In other words: one can provide the actor with resources (basic tactic) or one can make the actor provide himself with resources. A second alternative would be to influence the activities of other actors that are relevant to the resources and motivation of environmentally relevant activities of the target group. In other words: one can also make another actor provide the resources to the actor in the target group whose activities one wants to influence.

At closer look the two tactics described above have a multi-layer character. First, the policy actor tries to influence an activity that is not directly environmentally relevant. This activity could either be a form of flow of resources by the actor that can be seen as a member of the target group or it could be a (conditional) flow of resources between another actor and the target group. The four basic tactics could be used in this 'first step' by the policy actor. The result of this 'first step activity' (either by the target group or another actor) is a (conditional) change in resources and/or motivation that are related to environmentally relevant activities of the target group. The 'second step' in this tactic is thus the change in environmentally relevant activity due to the results of the 'first step activity'.

The first tactic could consist of two forms named '*self provision of resources*' and '*self deprivation of resources*'. As has been indicated above, this tactic involves the stimulation of activities by the target group that result in a change of resources and/or motivation that should result in a change of environmentally relevant activities by the actor. An example of such a tactic could be an obligation to make an environmental impact assessment for activities by the actor. The first step in this tactic can be seen as an obligation (basic tactic) to collect and produce information about the environmental consequences of alternative activities.[4] This additional information is a change in resources that may influence the choice between the environmentally relevant alternatives. If the information is made public it could also be used as a resource by the policy actor or other actors to perform additional activities to influence this choice of the actor in the target group. Another example might be the stimulation of the appointment of environmental managers in an organization. This 'change of stock in skilled people' might be stimulated by basic tactics like provision of information (convincing organizations that such a manager is to their benefit), conditional provision of money (a subsidy), providing skills to people (a training programme for managers) or deprivation of legal rights (an obligation to have an environmental manager). In the second step of this instrument the environmental manager might influence relevant activities of the organisation.

The second multi-layer tactic could consist of '*provision of resources by other*' and '*deprivation of resources by others*'. The first step in these tactics would be the stimulation of an activity of other actors than the target group. The policy actor might use the four basic tactics to stimulate this activity. The activity of the other actor would itself consist of the use of one or more of the basic tactics by the other actor directed at the target group. This activity would result in a change in resources and/or motivation in relation to the environmentally relevant activities of the target group. An example of this tactic is the provision of information to the target group by intermediate organizations. The first step might be the stimulation of an intermediate or interest organization to give information to the target group. Because this organization may have a high level of information on the target group and may have a high level of consensus/trust, this intermediate organization might be more successful in providing valued information to the target group than the policy actor that would provide information himself. A second example might be the formulation of a covenant between the policy actor and a branch organization. The establishment of this covenant might depend strongly on the exchange of information and the building of trust. The covenant may consist of agreements on activities by he branch organization to motivate actors in the target group (firms in the branch) to change their environmentally relevant activities. A third example of the exchange of resources between different actors is a deposit-refund system (for instance on bottles containing softdrinks). In such a system bottles and money are exchanged between consumers, shops and producers, preventing used bottles being brought into the environment, In the case that such a system would be the result of an activity of a policy actor (for instance by obliging producers to initiate such a system), one can see the system as a multi-layer policy instrument.

The examples of these multi-layer tactics given so far refer to a two-layer situation. Of course there may be many examples of three-layer tactics and of tactics having even more layers. One example might be the provision of a legal right to environmental groups to appeal to courts on the provision of legal rights to firms in a permit. The environmental organization can use this resource in a tactics of conditional deprivation of resources to the implementing agency (If you don't formulate strict standards in the permit we will take you to court), motivating the implementing agency to formulate strict standards that deprive the firm of the legal right to perform a polluting activity. Of course, as in all cases, the success of this tactics depends on the ability to create combinations of resources and motivation that are sufficient or necessary for the subsequent activities. In other words: the environmental group must have sufficient resources (people, money, time, etc.) and the motivation to actually use the right to appeal; the implementing agency must have the resources to formulate strict standards and the threat of appeal must be sufficient as a motivation to do so; and at last, the deprivation of the legal right to the firm must be sufficient to let the firm refrain from polluting activity.

Actor Organization and Reorganization as Instrumental Tactic

Until now four basic tactics have distinguished and two multi-layer instrumental tactics. These tactics depend on existing actors. As a final form of instrumental tactic one can distinguish the *(re)organization of actors*. In order to change the activities of existing actors a policy actor can create new actors (organization), eliminate existing actors or reorganize existing actors.

The *creation of actors* can consist of the creation of more or less permanent actors (like the creation of a new implementing agency for a policy) or can take the shape of organizations with a limited time horizon like 'working groups', 'task forces', 'negotiating committees' and 'research teams'. In order to function, these actors are to be provided with resources and some form of motivation. Provision of resources seems to be an important tactic in this respect. Sometime specific resources like a 'mediator' (skilled people) can be given to such an actor in order to produce specific resources like trust. Of course, other resources like money, skilled people, time, legal rights, etc., are also necessary for the activities of a (temporal) organization. A new organization can influence the activities of existing organizations by using the different instrumental tactics already discussed or by 'giving' resources and/or motivation in the form of members of the temporal organization that go back to the organization where they came from.

The *elimination of actors* can be seen as an ultimate solution to prevent further activity by that actor. Just as the death penalty can be seen as the ultimate possibility to prevent crime, the liquidation of organizations can be the ultimate solution to the problem of environmental degradation. An example of this drastic strategy can be seen in the Netherlands where the detection of high dioxin emission from waste incineration plants in 1989 resulted in the shutting down of four of these plants.

As a last instrumental tactic one can distinguish the *reorganization of actors*. Examples of this tactic are the shifting of tasks from one organization to another, the changing of coordination inside or between organizations and the changing of organizational structure. To some extent one might argue that reorganization is nothing more than an intensive form of changing resources and motivation either between or within organizations. With a reorganization however, there is generally a fundamental change in the nature and functioning of an organization that goes well beyond the reshuffling of some resources and/or motivation. The intensive nature of this tactic implies that it can generally only be used with organizations that are closely related to or are heavily depending on the policy actor (like implementing agents, research institutes of policy actors, etc.). The ultimate (environmental) effect of this tactic will depend on the change that it brings about in the (conditional) flow of resources to and from actors that perform environmentally relevant activities.

6. From Tactics to Instruments: The Classification of Policy Instruments

As has been indicated in the outset of this chapter, we were interested in a 'broad' and 'open' classification of policy instruments. As a consequence the result is not a full-grown 'matrix-shape' classification scheme with neatly exclusive boxes where concrete instruments can be fitted in. What is presented is a general method for the characterization of a concrete instrument in terms of the way it influences (or is supposed to influence) activities of actors. The reason that we have been speaking of 'instrumental *tactics*' and not of '*instruments*' is that concrete policy instruments seldom consist of just one form of what we called instrumental tactics. Often there is for instance a combination of providing a certain amount of different resources, together with conditional deprivation of other resources or even a combination with multi-layer tactics. We believe that the characterization of policy instruments in terms of these combinations would contribute more to the study of and discussion on policy instruments than a desperate attempt to fit complex phenomena like instruments in neatly exclusive boxes.

The actual classification (or characterization) of policy instruments can go along the lines of answering the following questions:

1. a) Does the policy instrument consist of elements that can be seen as an attempt to directly influence environmentally the relevant activities of the target group?
   b) If this is so, what basic instrumental tactics are used?
   c) What kind of resources are involved in these tactics?
   d) What kind of activities (flow of physical goods) are to be changed by the instrument?
2. a) Does the policy instrument consist of elements that can be seen as a multi-layer instrumental tactic?
   b) If this is so, what basic instrumental tactics are used in the first step?
   c) What kind of resources are involved in these 'first step' tactics?
   d) What kind of results are expected from the activity that is influenced by the first step tactics?

e) What kind of resources of the 'next-step' actor are involved in this instrument?
f) What kind of activities (flow of physical goods) of the ultimate actor (target group) are to be changed by the instrument?

3. a) Does the policy instrument consist of elements that can be seen as the (re)organization of actors?
b) If this is so, what kind of basic instrumental tactics are used for the (re)organization?
c) What kind of resources are involved in these tactics?
d) What kind of activities (flow of resources) are to be changed by the instrument?
e) What are the expected results for the ultimate actor (target group)?

In order to give a rough understanding of the way in which policy instruments can be characterized by this conceptual model, it seems instructive to give some examples.

A *charge system* could generally be seen as conditional deprivation of money. As an example one could think of the Dutch tax differentiation on cars. The amount of the tax (the charge) is dependent on the cars' emission of CO, HC and NO_x. The tax difference between cars that meet certain standards and cars that don't, functions as a motivator for the car importers to use catalytic converters (the extra costs of the catalytic converter are covered by the lower amount of the tax to be paid). The charge system however also deprives the car importer of certain legal rights. Once they sell a car model as meeting the standards for the low-tax category, they have to make sure that all cars of this model meet these standards (legal obligation). The car importer can even be obliged to take all cars of a model back for retrofitting if the control-programme shows that cars of this model usually don't meet the standards.

Eventually this charge system is not directed at the behaviour of the car importer, but at consumers buying cars. The tax system aims at creating a supply of clean cars that is sufficiently attractive to be bought by consumers (a multi-level tactic). In order to stimulate the sale of cars with catalytic converters, government accompanied the tax differentiation with a persuasive media campaign, trying to motivate consumers to buy these cars by providing information. This basic tactic also functions as a multi-level tactic influencing the car importers. By supplying cars with catalytic converters the importers could develop a 'green image', improving their trust in relation to both consumers (who were motivated to buy 'green cars') and government (who wanted 'green cars' for environmental purposes).

Another illustrative example might be the *covenant* as a policy instrument for expressing mutual agreement between the policy actor and (representative organizations of) the target group. The formulation of a covenant can generally be seen as a process of creation of mutual trust by developing some sort of consensus. this process mostly takes place in direct negotiations (exchange of information) between representatives (skilled people) of the policy actor and the target group. As an example one could think of the Dutch covenant on the banning of CFs in spray-cans. This covenant contains an expression of mutual trust, where the policy actor and the target group state their shared commitment to the reduction of the use of CFs. It is also stated that agreements they had made in the past had been successful and were met by the target group. Rules were set on the production of information by the target group and the processing of data by an accountancy firm. This information functions as a detector for the policy actor in order to make clear whether the reduction programme was met by the target group. For the target group the information could be used to monitor their reduction programme (a multi-level tactic). It was agreed that the cost of data processing was to be paid by the policy actor (provision of money). It was also agreed that firms that would have to move their production facilities in order to be able to use the alternative gases would receive financial support from the policy actor (conditional provision of money). For one firm it was eventually agreed (after consultation of local government) that the use of alternative gasses was allowed (provision of legal rights) in order to prevent the moving of the production facility.As can be seen in this case, the policy instrument (the covenant) contains a number of different instrumental tactics concerning a number of different resources.

Notes

1 The classification of resources that is presented here is an adaption of the classification of the resources of a policy that is present in: Knoepfel, Peter, unter Mitarbeit von Rita Imhof, *Oekologische Vernetzung versus Rechtsstaatliche Handlungmaximen – Möglichkeiten zur Überwindung eines gespannten Grundverhätnisses*, Lausanne, 1991.

2 If time is a resource, one might also conceptualise 'space' as a resource. In the model developed here, space is seen as a physical good and consequently not as a resource of its own.

3 Compare Benson, who distinguishes two resources: money and authority (Benson, 1978).

4 An obligation can be seen as the deprivation of the right to refrain from an activity.

References

Benson, J.K. (1978) The interorganizational network as a political economy, in L. Karpik (ed), *Organization and Environment*, London, pp. 69–101.

Doelen, F.C.J. van der (1989) *Beleidsinstrumenten en energiebesparing* (diss.), Enschede.

Hood, C. C. (1983) *The Tools of Government*, London.

Klok, P.-J. (1991) *Een instrumententheorie voor milieubeleid* (diss.), Enschede.

3. The Political (Un)feasibility of Environmental Economic Instruments*

CORINNE LARRUE
Centre d'Etudes Supérieures d'Aménagement, Tours, France

1. Introduction

Throughout this paper we will address the issue of the political (un)feasibility of economic instruments, i.e., the manifold obstacles to an economization of environmental policies. This question is linked with several issues already addressed in the previous chapters: the definition and classification of policy instruments (P.J. Klock) and the question of arguments behind the choice of policy instruments (A. Liberatore).

Usually, *economic instruments* are defined as instruments which tend to influence the ratio between individual costs and benefits of several options. They are based on the polluter-pays principle which states – as defined by O.E.C.D. – that the polluter should bear the cost of measures to reduce pollution decided upon by public authorities to ensure that the environment is in an "acceptable state" (OECD, 1986). They contrast with "command and control" instruments, which tend to directly influence polluter behaviour, and with persuasive one, which influence the individual or collective perception of the costs and benefits of different options (OECD, 1989). Last, services directly provided by the public sector can be seen as another policy instrument (Maintz, 1979), the implementation of which is linked with economization of public policies. In the field of the environment, such instruments play an important role for waste water utilities for instance.

The survey of economic instruments for environmental protection made in 1989 by O.E.C.D. pointed out several types economic instruments in the field of environmental policy (OECD, 1989):

- *charges*, i.e., a burden to be paid by polluters or consumers; they regroup effluent charges (to be paid on discharges into environment), user charges (to be paid for the cost of collective treatment of effluents or for the private usage of natural resources), product charges (laid upon the price of polluting products), tax differentiation (which lead to more favourable prices for envi-

Bruno Dente (ed), Environmental Policy in Search of New Instruments, 37–54.

ronmental-friendly products or processes), and administrative charges (to be paid for registration or enforcement of law);

- *subsidies*, i.e., a financial assistance provided to polluters by means of grants (non-repayable) soft loans (repayable but with low interest rate) or tax allowances (accelerated depreciation of anti-pollution investments);
- *deposit refund systems*, i.e., a surcharge laid on the price of potentially polluting products refunded when the product is adequately treated;
- *market creation*, i.e., artificial markets created in order to let polluters exchange their right to pollute (emissions trading system) or their recycled materials (by means of price intervention); legal liability insurance of polluters for environmental damages (which induce an insurance market) can also be put into the same category;
- *financial enforcement incentives*, i.e., fine to be paid for non-compliance behaviour (non-compliance fees) or payment returnable upon compliance (performance bonds).

This paper will focus mainly on charges (effluent, user and product charges) and, though to a lesser extent, on market creation (emission trading), which can be considered as the main economic instruments in operation in the field of environmental policy. From a theoretical point of view, these economic instruments share a common incitative element, which justifies the choice made. Nevertheless this choice led us to speak mainly of *economization* of environmental policies instead of feasibility of economic instruments.

By *political feasibility* we mean the ability to translate these economic mechanisms into formal rules. Economic instruments need to be formalized in order to be implemented. This formalization requires a minimum of support from institutional actors and from target groups. This paper tries to show why this formalization is scarcely supported. As a matter of fact, despite the support stemming from economists and despite recent progress, economic instruments are not as widely used as it can be wished for.

This paper is partly based upon research we have carried out on implementation of environmental policies in several countries and for different fields (Larrue, 1988a and 1988b, 1990; Knoepfel and Larrue, 1985, Knoepfel, Larrue and Zimmerman, 1987); and specially upon the very first results of research undertaken on the cost and financing of water and waste water utilities, in the United States, France and Germany (Larrue and Navarre, 1990). It is not strictly based upon empirical findings but on first thoughts which came out during these research efforts; it is also based on the literature, mainly O.E.C.D.' works on economic instruments (OECD, 1989a and 1989b, 1991); it has been considerably enriched by the comments from the ESF Task force members.

After having briefly defined (2) what economization of environmental policies means, we will consider three basic factors which seem to play a role in the shortcoming of economic instruments. The first one (3) called procedural obstacles, is linked mainly with budgetary principles and with the complexity of such a process; then (4) institutional obstacles will be pointed out as being a determinant. Last (5) we will focus on more structural obstacles, which explain the difficulties in finding support for an economization of environmental policies, from public actors as well as from target groups.

However for each one of these factors, we will try to set out, on the one hand (negative) arguments which explain the relatively poor success of economic instruments, and on the other hand (positive) arguments which have been in favour of economic instruments (for those already in operation).

2. About Economization of Environmental Policies

An understanding of this paper will require at least some information about the concept of economic instrument and more generally about economization of environmental policies. Nevertheless, we do not intend to spend much time specifying a new – but still controversial – definition of the economic instrument, which has been attempted many times; we want only to clarify the concept of economization of environmental policies as referred in the text.

As pointed out above, this paper will focus on few main economic instruments, namely those of charges (effluent, user and product charges) and market-creation (emission trading).

Charges are often considered as a "price" to be paid for pollution: "polluters have to pay for their implicit claim on environmental 'services' (OECD, 1989, p. 14). Various charges can be identified: effluent charges, to be paid on discharges into the environment are the best known, as are user charges, to be paid for the cost for collective or public treatment of effluents, or for water delivery. Others are less developed, like product charges, laid upon the price of polluting products, and their counterpart tax differentiations, both of which lead to more favourable prices for "environmental-friendly" products.

From an economic theoretical points of view, *charges* must have an incentive function. Base and rate of effluent or of product charges might be fixed in such way that they will lead the polluters to internalize the cost of the pollution, i.e., the polluters will include in their production costs the cost of the damages caused by their pollution. For this reason, the charge base must be linked with the quantity of pollution actually discharged by polluters or induced by the product, and the rate of the charge must be set at an optimal level, entailing an optimal level of pollution. An optimal level of pollution is one for which (marginal) cost for purification or non-pollution equalizes the (marginal) cost of pollution (i.e., the marginal cost of damages). For such a level, the collectivity will not devote too many resources to the protection of the environment but will do it in sufficient quantities.

The same is true for user fees. User fees are used mainly in the field of water delivery, industrial and domestic waste water utilities, and for the collection and the treatment of municipal and in some cases industrial solid waste. From an economic theoretical point of view, in order to achieve an optimal allocation of (public) goods, and thereby in order to protect the environment, the price of user fees must be set at a level which corresponds to the marginal cost of the service provided, i.e., the cost induced by an additional demand of water, or of waste water and solid waste collection and treatment. In such a case, one can demonstrate that the collective utility is optimal.

In the case of *market creation* instruments artificial market are created where individual economic actors can buy and sell their "pollution rights", the main form of which is emissions trading. Such instruments rely upon a previous formal and legal definition of pollution rights. It (pre-)fixes the overall limited amount of pollution to be emitted and allows polluters the choice of means to comply with their limited quota (possibly by means of quotas trading).

Charges and emission trading do not fulfil the same purpose: effluent charges, product charges and user fees tend to internalize pollution costs, whereas emission trading tend to reduce the cost of pollution abatement. However, they both come under economic instruments family, since they introduce market mechanisms as an instrument of environmental policy.

As concern the economic instruments presently in operation, one can notice the existing difference between the United States' way of economization of environmental policy, and that of most European countries. As a matter of fact, if the Federal and State governments of United States seem to be reluctant to use charges (whether effluents charges, product charges or user fees), they seem to be more comfortable with marketable permits (OECD, 1989). Whereas European countries share in common the use of charges, particularly effluents charges, very few European countries have implemented emission trading policies (OECD, 1989). This paper will try to point out the reason for such a discrepancy. Note moreover that the United States started much later: pollution charges were introduced in Europe in the sixties whereas the American emission trading mechanism was introduced in the late seventies.

More generally, the purpose of this paper is not to demonstrate either the theoretical effectivity of economic instruments or the weak or misuse of these instruments. We take for granted the fact that from a theoretical economic point of view, the economic instruments defined above will have an incentive impact on polluters' behaviour, provided they are set at the optimal level. We also take for granted the fact that, presently, in many countries, very few economic instruments are implemented in the field of environmental policies, compared with the so-called 'command and control' ones. As pointed out by O.E.C.D. works, economic instruments 'remained secondary to other instruments' (OECD, 1989, p. 9) and recommendations are made at higher level toward a more systematic use of such instruments (OECD, 1991, p. 9).

However, if 20 years ago economic instruments met the opposition of many policy actors – predominantly ecologists – today their implementation is less controversial. More and more environmental economic instruments are being utilized by governments, probably because command and control instruments failed to improve the environment. But their formulation does not really follow economists' recommendations. Many so-called economic instruments implemented in the field of environmental policies do not comply with several constituent criteria of economic instruments. For instance, charges generally serve more as fund-raising instruments, even when they were designed to have an incentive function (and if, all things equal, they will have one).

So, the paper will focus on those cases where any economic instrument is used, and where those economic instruments do not play the role previously expected.

The reasons why economic instruments lack support vary from country to country, and from one kind of instrument to another. We will focus on the main factors behind the ineffective employment, or even outright misuse, of these policy instruments, which can be met partly or entirely within a single country, and which may apply to one or more kinds of economic instrument. For that reason, our review of the obstacles of the economization of environmental policies cannot be considered an exhaustive one.

Finally, we are aware that most of the arguments presented below apply also to non-economic instruments. However the question addressed here is not 'why command and control instruments are preferred to economic ones', but 'what are the main obstacles to economic instruments', even if these obstacles are an issue for both kind of instruments.

3. Procedural Obstacles to an Economization of Environmental Policies

Two main obstacles can be stressed which apply mainly to charges, and to a lesser extent to emission trading. The first one deals with budgetary process and the second with the complexity of the process.

(a) *Budgetary Obstacles*

Economic instruments involve financial practices, i.e., taxation charges, and price setting. In many cases these financial processes, when applied in the field of environmental policy, do not fit within the general framework of the budgetary process. Economization of environmental policies suffers from a lack of formal resources (to refer to P.J. Klock's classification). Utility user charges, in the case of the United States, provide a good example of such an obstacle.

In the United States (except in California) user fees for water delivery and for waste-water utilities do not fulfil economic requirements. They are seldom or even never set at the marginal cost level. Moreover, in some cases they work in exactly the opposite way, inasmuch as they introduce declining block-type rates for water consumption and, though to a lesser extent, for waste-water treatment (Boland, 1983).

The main argument for such a rate-setting process, given by local authorities in charge with the management of these utilities, is linked with budgetary rules. In the United States, local governments must set rates only at a level which ensures a budgetary balance: they are not allowed by law to raise more revenues than those sufficient to recover operating expenses, debt service and capital outlay requirements. In this case the principle of marginal cost pricing do not comply with budgetary principles. One may notice that a practical consequence of such a discrepancy is that the local agencies responsible for the management of these utilities cannot make anticipatory provisions for the rebuilding of networks for instance, and have to borrow money case by case; the quality of utilities networks suffers from these financial practices. The same applies to effluent charges, which cannot be introduced at this administrative level in this country.

This general budgetary framework which requires the government to abide by, acts also (and especially) against charges which stand below operating and capital expenditures. The incentive impact amounts to charges, which will have an effect upon the amount of money collected and which is difficult to manage from a financial point of view. Usually, taxes are set to raise revenue in order to finance public expenditures, the classical functions of which are allocating, distributive, and stabilizing – to refer to Musgrave's classical definition of the functions of the public budget (Musgrave, 1959). Then levies with a targeted incentive function – which nevertheless participate to ensure the allocative function – are more than marginal compared with the necessity to finance public expenditures.

More generally, uncertainty about the amount of money collected (in whatever way it is collected), which is a component of the economization of environmental policy, does not fit in with the general rules of budgetary process. Economic instruments do not accord with the principle of neutrality[1] which is one of the main principles for budget setting, even if this 'non-neutrality' is precisely the targeted effect. This uncertainty argument acts as an obstacle for the governmental ministry responsible for the budget, which do not easily accept any kind of economic instruments. But it will affect also the ministry responsible for the Environment, which generally needs the money collected to subsidize its ecological investments. Consequently, the environmental ministry will tend to support only those economic instruments capable of raising the most revenue, i.e., it will not pay attention to their incentive impact.

However, one way was found to bypass this problem, *viz.* the creation of *ad hoc* committees charged with the management of these economic instruments. In France and the Netherlands, for instance, the charges imposed on effluent discharge into water are set and collected by special water agencies created just for this purpose. Moreover, this obstacle can be overheard within a traditional administrative framework as in Germany, where the *Länder* have been able to collect since 1981, effluent charges with strong incentive effects. One can conclude that this procedural obstacle needs organizational resources to be overcome.

(b) *A Complex Process*

The second procedural obstacle is linked with the complexity of such a rate-setting process. To refer to P.J. Klock's classification, one can say that the economization of environmental policy suffers from a lack of knowledge. The complexity arises from the fact that such charges or prices have to be set at an optimal level, which is not easy to determine in practical terms. This is true for user fees and, above all, for effluent charges and product charges.

Agricultural pollution of water by phosphates and nitrates gives a good example of such complexity. Nitrate and phosphate pollution is a result of slurry spreading and certain fertilization practices. It is more or less easy to determine charges on husbandry activities in order to tackle the slurry-spreading problem: a tax imposed per head of livestock over a given amount of cattle per acre (even taking into account the quality of the soil) is easily enough defined (as has been done in the Netherlands). But it is much more difficult to define precisely the base and level of a charge laid on crop farming. We do not know any accurate way of measuring the amount of nitrates present in the soil for any given farm or field, and even if we did know such a means, the administrative cost of its enforcement (with inspection and testing costs) would be prohibitive. On the other hand, levels of pollution stemming from crop farming activities will vary from one field to another even when same amount of fertilizer has been used (due to differences in the quality of soil), and also from one farmer to another (due to the farmer's skills). In this way a simple tax laid on fertilizer consumption fails to take all factors into account and so would appear less effective than the targeted-incentive approach. This complexity is an important reason behind the present difficulties in laying charges on agricultural activities in France and United Kingdom, for instance.

More generally, two kinds of difficulty appear here. The first one deals with the determination of the optimal base and rate of the charge. As concerns the base of the charges, depending on which discharge or product you consider, it can be very difficult to isolate adequately a pollutant which is at one and the same time identifiable (i.e., quantifiable) but also representative of the sort of pollution targeted. The base must be clear enough to have an incentive impact and also enforceable (i.e., not entailing prohibitive administrative costs) to have a chance of succeeding once it has been implemented; this can be very complicated for discharges into atmospheric environment for instance (OECD, 1991). As concerns the rate of the tax, we have seen already that it has to be fixed at near the optimal level in order to have an incentive impact balanced with the potential polluting impact. This optimal level is particularly hard to define by a public administration. On one hand, a public administration will not immediately know the cost of effluent purification for the manifold industries involved, and thus will have difficulty in setting prices or charges at an adequate level; on the other hand, even if such information were available, one can assume that in practice, an incentive impact can be met at a lower level than one theoretically determined. As a matter of fact companies do not know precisely the level of charges which underlies their purification costs, and will probably be inclined to depollute by a charge the level of which is lower than a theoretical one. Then to be effective, administration must proceed by trial and error, an attitude which does not fit with a classical way of public action.

Moreover, with user fees, an additional complexity appears, which relies upon the choice of marginal cost to be taken into account. A choice can be made between short-term marginal cost, which includes operating and maintenance costs, and long-term marginal cost, which includes capital expenditure, choice upon which economists do not all agree.

The second difficulty stems from the fact that to be effective the rate of the tax must take into account the quality of the media into which pollutants are discharged. This condition implies that charges will vary from one region to another and from time to time. It is in this way impossible, or at least very difficult, to set fixed prices or stable ones. This instability is undoubtedly a factor in there being a weak incentive value intrinsic in the charges presently in operation in European countries. It is easier to set charges at a level which raises a given amount of revenue, than to set them at the optimal level as regards the discharges of pollution.

One can notice that this complexity underlies also the control induced by such instruments. The implementation of charges requires continuous information about emissions of each polluter and about the effectivity of purification plants. Such a control framework can be very heavy, entailing prohibitive costs.

The same is true for emission-trading processes. The rules for exchanging quotas between firms might be very difficult to define, and the control over the actual discharges of each company into the environment requires a lot of power and staff resources, not often available. U.S. experiments of such instruments in the field of air pollution shows that the implementation of emission trading entails important administrative works (OECD, 1989).

Nevertheless, an argument can be used to tackle these procedural barriers. It deals with the staff capacity of the environmental administration. Usually civil servants belonging to environmental administrations are engineers or jurists, i.e., not very familiar with economic mechanisms. But more and more economists able to manage such instruments are becoming involved within these administrations, which partly explains the growing importance of the economization of environmental policy.

4. Institutional Obstacles to an Economization of Environmental Policies

Institutional aspects deal with actors' involvement. Obstacles to an economization of environmental policy stem mainly from the fact that many actors are involved within the formulation and the implementation of economic instruments, and this tends to complicate the process. Participation conflicts may be located at the *horizontal* level (conflicts between the ministries of the central government) but also at the *vertical* one (conflictual relations between local, regional and national governments and also with international bodies). Moreover economic instruments suffer from a lack of support from lobbies and from target groups.

(a) *Conflicts Between the Ministries of the Central Government*

As a matter of fact, economic instruments do not pertain exclusively to environmental policy. Their formulation encroaches upon the hinterland of others ministries, mainly those responsible for the economy and for the budget ones. This institutional argument is connected with the one developed by Knoepfel, about the implementation of environmental policies by means of inter-policy cooperation.

The minister for the environment has to negotiate his/her own instruments in order to implement them. This negotiation process goes far beyond the classical negotiation process of between ministries for their budgetary allocations, which occurs each year and for each ministry. As a matter of fact, the minister for the environment must 'borrow' some of the legitimacy of the budget or economy ministries.

The involvement of other actors at this central level hampers the economization of environmental policies in two ways. First of all, because of their financial dimension, economic instruments must not only fit with environmental policy objectives, but also with more general financial or budgetary ones. Restrictions imposed by fiscal pressure and, more generally, cuts in public expenditure, are one of these financial general objectives. Effluent or product charges such as user fees will come into conflict with financial constraints inasmuch as they will potentially increase fiscal pressure. Thus one can say that a conflict occurs between global objectives and sectorial ones – to use Jobert and Muller's terminology (Jobert and Muller, 1987) – which is at odds with an economization of environmental policy.

Such a conflict explains the fact that it was easier for the U.S. Environmental Protection Agency to propose emission trading systems instead of effluent or product charges, because the former did not come into conflict with the general and very important opposition toward an increase of public revenue raising. To that respect, institutional factors not only tend to explain the limited implementation of economic instruments but also the choice among them.

Another conflict which falls into this category is the one between the struggle of inflation against effluent and product charges. In France, for instance, the recent National Plan for Environment (Chabason and Theys, 1990), proposed a set of effluent charges (on air pollution, waste production, noise produced by aircrafts or one urban impermeable areas, etc.), and product charges (tax on fertilizers). Most of these proposals were cancelled after being negotiated with the Ministry of the Budget, mainly for anti-inflation reasons. Here again this factor explains as much the absence of economic instruments as the low incentive level of charges when finally accepted. It is well known for instance, that, since their creation, an increase of effluent charges on waste water discharges in France was – and still is – limited by Budget ministry, against the will of the Minister for the Environment. Likewise, such an obstacle explains the choice for using non-economic instruments, i.e., instruments which do not reach into hinterlands of other ministries. For instance, one may note that in many cases, and especially in Germany and the E.E.C., a 'labelization' of products has been preferred to tax differentiation.

The second way in which the economization of environmental policies is hampered by institutional obstacles arises from the fact that usually the power of the ministry for the environment is very low when compared with that of the ministry responsible for the economy and Finance. Within the negotiation process between the manifold public interests, objectives supported by the budget easily win over sectorial ones. Consequently, ministries for the environment generally favour instruments falling fully within their own jurisdiction. This unbalanced situation has two main consequences: the first one is that in many cases, ministries for the environment will themselves seldom support economic instruments; the second one, is that, in a case where a ministry of finances would accept an economic instruments for environmental protection, it will more likely support the fiscal objective behind it than the incentive objective. Thus 'uncompromised' economic instruments have difficulties in finding supporters.

Nevertheless for a few years, the position of the Minister for the Environment has been strengthened by international pressure, originating mainly from E.E.C. In the wave of European Community building, each provision enacted in order to comply with E.C. requirements will have the benefit of more power. The balance between sectorial and general objectives tends more and more to favour environment, even if such an argument does not yet benefit economic instruments.

(b) *Conflictual Relations Between the Different Levels of Government*

Institutional obstacles also influence the definition of an optimal level of government able to formulate and to implement economic instruments. In this respect, the central government appears to be in many cases the best decision level (Prud'homme, 1984).

First of all, as seen in the frame of procedural obstacles, in many cases local governments are not empowered to lay down new levies. In many countries, local finances are controlled by central government and/or by law. This usually forbids the creation of a new tax or charges, even in order to protect the environment.

Secondly, even where local governments would be allowed to use financial means, they will not have any incentive to introduce economic instruments inasmuch as they and the inhabitants or activities under their jurisdiction will bear the total cost of the implementation of any such instrument, and will only partly benefit from it. Incentive rates for user fees, effluent charges or product charges created at a local level will be paid by local inhabitants or companies, whereas a cleaner environment will be of benefit to the regional or national surroundings. Moreover, if one region introduces effluent charges for its own territory, without being followed by the other regions, it runs the risk of seriously handicapping itself with respect to the competition for employment, which presently occurs between localities. This argument applies also to user fees or product charges. Consequently, local government will never or only seldom introduce such economic instruments on their own.

Moreover, local government representatives will be opposed to such an introduction by the central government. The case of the new water law in France is a good example. Following the recommendations of the National Plan for the Environment, the governmental project for a new water law, recently presented to the French parliament, proposed to lay a tax on urban impermeable areas, in order to incite local government to use more permeable material for building roads and/or to leave more free (permeable) areas within cities. The money collected would have been used to subsidize rainfall water treatment plants. But when the law was discussed within the parliament, the proposal was cancelled by the representatives of all the major French cities, which had been directly targeted by this financial provision.

The competition argument proves true also for the national decision process. Generally, the international competition stems from industries against economic instruments – and more generally against any anti-pollution provisions. However such an argument falls down when requirements are the same for each E.E.C. member state, for instance. Then in many cases an international decision process is preferable to a national one, inasmuch as it balances market conditions. But, if theoretically such as international level is suitable, one can notice that the decision process is complicated enough at the E.C. level, as it involves more partners than at a national level. The case of CO_2 tax provides a good example of such a complexity (cf. Liberatore's paper). As a matter of fact, one can assume that many difficulties will be experienced in establishing of this tax.

(c) *Lack of Support from Lobbies and Target Groups*

Another element to be pointed out here is the little support of economic instruments stemming from ecologist organizations. Until recently, ecologists considered that economic instruments amounted to buying the 'rights to pollute', which was in opposition with their own environmental protection philosophy, where everybody is responsible for their actions with respect to the environment. However, their position changed over the time, and economic instruments now carry more credibility within environmental organizations. The case of Clean Air Act of 1990 in the U.S.A. provides a good example of such change. Within the negotiation process, the environmental lobby did accept the emission trading principle, provided that the overall amount of SO_2 emissions was limited (personal communication from an E.P.A. official).

Nevertheless, as concerns charges, such support does not apply to the incentive function of economic instruments, but much more to their revenue-raising function. Economic instruments found support among ecologist lobbies mainly because they provided a financial means to environmental policy. Nonetheless, the economic requirements underlying this kind of instrument will not find any support from these organizations any longer.

Likewise, economic instruments lack support from target groups, mainly industries and households. Such a position is easily understandable as these two economic agents bear directly the burden of the charge. Moreover, industrialists seem to prefer 'command and control' instruments as they consider them to be more power within the negotiation process. Moreover, as Buchanan and Tullock have argued (Buchanan and Tullock, 1975), 'command and control' regulation leads to barriers to entry, and hence to higher existing polluters profits.

However, the industrial anti-polluting sector, which will benefit from a strongest environmental policy tends to play a lobby role, pushing such instruments to the front of the policy arena. In many countries an 'eco-sector' is growing rapidly. Such a sector will benefit from severe standards, but also from economic instruments, which lead to more competitive antipollution equipment. To refer again to P.J. Klock's classification, economic instruments will benefit more and more from new motivation.

The lack of support from these different actors can be explained by structural obstacles to an economization of environmental policy.

5. Structural Obstacles to an Economization of Environmental Policies

Three main elements can be pointed to as being structural obstacles to an economization of environmental policies. The first one deals with the distributive impact of economic instruments, the second with the perception of this distributive impact and the last one more generally with the role of the State within our developed society.

(a) *The Distributive Impact of Economic Instruments*

The distributive impact of economic instruments refers to the question of who will bear the burden of the costs for the protection of the environment and who will benefit from it. This question applies also to environmental policy itself; economic instruments just show explicitly the burden of the costs and the recipients of environmental protection than regulatory ones.

Generally speaking, one can say that environmental protection entails a cost to be borne either by the polluted, by consumers (of goods produced by polluters), by taxpayers (when States bear the burden), or by the three of them in combination.

A shift to economic instruments usually changes the distribution of the burden among the groups having to bear it, as compared with a distribution induced by 'command and control' instruments. For instance, effluent charges will usually transfer a part of the burden from the polluted group to the polluter group, inasmuch as the latter will add the burden of the charge to its purification costs (Prud'homme, 1981). As a matter of fact, the 'command and control' instrument means the polluted bear some of the pollution costs, i.e., the cost of the pollution authorized by the emission standard, whereas effluent charges lead to internalize of all the pollution costs. Such a transfer explains the opposition stemming from industrialists who generally prefer 'command and control' instruments as they are costless for them.

Moreover, the higher production costs induced by charges will generally be translated into consumer prices, in such a way that the consumer will bear the additional costs. Then the question of social equity arises. A distribution which lays the burden upon taxpayers – which comes into conflict with the polluter-pays principle – can be more equitable than a distribution based on consumption, provided that the fiscal system is progressive (i.e., when taxpayers pay in proportion to their revenues). Such an argument prevents governments supporting economic instruments.

The incidence issue concerns also spatial distribution. For instance, any charges laid on the transportation sector in order to internalize environmental cost will benefit the central regions (where transportation costs are minimized), and will handicap peripheral ones. In such a framework, environmental protection will come into conflict with spatial planning, for instance. Economic instruments which tend to make more visible the burden of the environmental costs emphasize these conflicts.

Many arguments have already been written on this issue. Among them, the O.E.C.D. recommendations about the implementation of economic instruments (OECD, 1991) pointed out the provisions to be made in order to reduce any negative incentive stemming from the distributive issue. It concerns mainly the financial compensations to be introduced in favour of the injured parties. Such forms of compensation have already been implemented in countries like France, in the case of waste-water effluent charges. As a matter of fact, in this country, financial-aid systems are closely linked to charges systems. The revenue raised by effluent charges are redistributed to those firms which bear them, in order to incite investments in purification plants. Such a mutualistic system was a strong argument which led industrialists into accepting this effluent charges system.

O.E.C.D. recommendations concern also the information to be provided before and during the decision process, about the benefits surrounding the use of economic instruments for environmental protection, but also for target groups. As pointed out by O.E.C.D. studies, support from target groups is essential to allow an implementation of economic instruments. Target groups must then be convinced of the profits they will get from such instruments (mainly flexibility).

(b) *The Perception of the Distributive Impacts*

The issue of who will bear the costs and the benefits is not the only one to be addressed here. An additional question will be asked concerning who is perceived as bearing the costs of and the benefits from these instruments. The obstacle is as much linked with the visibility of the distributive impact of economic instruments as with the actual distribution of costs and benefits.

The benefits of environmental protection are diffuse in spatial and demographic terms. The improvement of the environment will be of benefit to the population on a whole, whereas the costs will be concentrated on a few economic agents, especially when economic instruments are used. Moreover, the benefits of environmental protection appear to be diffuse also on temporal terms. In many cases, an effort made today will only profit the next generation, and this does not constitute an incentive to the present one.

In such a configuration, one can say, with Padioleau in his attempt to classify regulatory policies (Padioleau, 1982), that the formulation and the implementation of environmental policy is highly conflictual, and the probability of success is very low. One can state that the difficulties in implementing environmental regulatory policies (which profit from a limited but existing support) will be emphasized as concerns economization of environmental policies which do not profit from any support from target groups. In a controversial field, any additional difficulty will increase the shortcomings of the policy.

More generally, the profits derived from the economization of environmental policies and from the environmental policy itself are not often easy to discern. The protection of environment requires financial outlays from many distinct economic sources. These individual efforts are tangible, while the collective benefit is rather abstract. For individuals the balance appears not to be fair. As a matter of fact, the sum of individually realized benefits will not equalize the actual collective utility. The decision process in our democracies relies upon the will of the majority and it is clear that this majority will not generally be in favour of economic instruments. There is a strong opposition between an egalitarian political system and the inegalitarian economic system.

The case of the LASER project set up in order to tackle the traffic congestion problems of Paris provides a good example. The LASER project proposed building an underground road which would have crossed Paris from one side to another. The price for using this underground road would have been very high. An economic evaluation showed a collective benefit of such a project, as well as a benefit for environment (Papon, 1991). Time lost to traffic congestion costs a lot to the collectivity, and these costs would be recouped by the LASER project. Moreover, such a project would have a positive impact on the quality of the social environment and on air pollution on the street of Paris (both by reduction of traffic congestion). But the project tell victim to the perception of being elitist, of benefit only to the rich, and was cancelled in the face of the opposition, stemming from the main part of the population but also from ecologist groups. In this case, the perceived individual profit appeared to be less important than the collective one.

(c) *A New Role for the State*

Structural obstacles rely generally upon the role of the State within our developed society. Since the last world war, the role of the State has become mainly a distributive and an allocative one, known as the Welfare State, the main means of which are public finances. Government uses public funds to produce goods and services which are not or cannot be produced optimally by market mechanisms and to transfer large amounts to 'deserving' groups. Even taking into account the latest wave of liberalism which has for the last decade tended to reduce the level of State intervention, most public expenditure still seeks to provide public goods and to redistribute revenues in social as well as geographical terms.

Compared with such a role, the economization of environmental policy works entirely in the opposite way, inasmuch as it creates scarcity (before it really exists). As we have already seen, user effluent fees and product charges represent – or might represent – the price to be paid for the environment. Likewise, an emissions trading system supposes that 'environmental consumption' is previously and artificially limited. The environment thus becomes an artificially scare resource, whose scarcity has to be factored into production costs. Such a role does not fit in with the classical way of public action, based on the distribution or redistribution of goods. It presents a new way of taking into account public interests, which goes against the single and visible economic interest of inhabitants and activities. This new way of action is presently supported by the ideological movement toward so-called 'sustainable development', which propounds the voluntary limitation of human activities. In order to bring about sustainable development its promoters propose the adoption of the principle that our chances of development must not influence those of the next generation. Such an attitude will hardly fit into the general framework of the political and economic decision making processes.

6. Conclusion: Toward an Economization of Environmental Policies?

These many obstacles to an economization of environmental policies, as stated above, must not discourage the promoter of economic instruments. Indeed, the one conclusion that can be derived from our paper is that the economization of environmental policies is not an easy task; but it must not be considered as an impossible one. However, promoters of economic instruments must not assume that economic instruments are theoretically efficient – even more efficient than command and control ones – they can easily be translated into political decision process. As shown above, environmental economic instruments go so much against the 'natural' way of public action, that they require substantial change to overcome this opposition. In that respect, we can wonder if the main obstacle to an economization of environmental policies is not the economists themselves inasmuch as they usually do not anticipate such natural opposition.

The promotion of economic instruments requires then the development of information resources aimed at convincing the manifold actors involved, including and mainly target groups. In that respect, the O.E.C.D. studies constitute a first step in this direction, which might be followed on the national level.

However, a second conclusion may be drawn. It is that whatever the promotion of economization of environmental policies, economic instruments will unavoidably be altered by the political decision process. This alteration will affect mainly the incentive-creating value of economic instruments. As the history of environmental policy shows, agreement is easier to obtain on the revenue-raising function of charges than on their incentive function. Likewise, if emission-trading systems suppose an explicit and previous limitation of overall emissions, they might either not be supported by the target groups, or not be implemented as such. Thus it will help the adoption of economic instrument if economists work not only with the theoretical impact of economic instruments, but also with the effects of politically altered economic instruments.

Notes

* Paper written for the European Science Foundation Programme on Environment, Science and Society; Task Force on Policy Instruments. Thanks to R. Prud'homme, R. Darbéra and all the members of the Task Force for their comments.

[1] This principle states that a good fiscal system might not impact the allocation of goods.

References

Boland, J. (1983) *Water/wastewater Pricing and Financial Practices in the United States*, Technical Report 1, MetaMetrics Inc./Agency for International Development, Washington.

Buchanan, J.M. and Tullock, G. (1975) Polluter profits and political responses. Direct control versus taxes, *The American Economic Review* **65**, 139–147.

Chabason, L. and Theys, J. (1990) *Plan National pour l'Environnement, Rapport Préliminaire en vue du Débat d'orientation*, Ministère de l'Environnement, Paris.

Jobert, B. and Muller, P. (1987) *L'Etat en Action. Politiques Publiques et Corporatismes*, Presse Universitaire de France, Coll. Recherches Politiques, Paris.

Knoepfel, P. and Larrue, C. (1985) Distribution spatiale et mise en oeuvre d'une politique publique: le cas de la pollution atmosphérique, *Politiques et Management Public* **3**(2) 43–49.

Knoepfel, P., Larrue, C. and Zimmerman, W. (1987) Regulatory change and institutional rearrangement: building new policy arenas for ecologization of agriculture. A comparative analysis of programmes in France and Switzerland, *Social Science Information* **26**(4).

Larrue, C. (1988a) *La Pollution des Eaux d'Origine Agricole: Comportements Agricoles et Politique Publique. Deux Études de cas dans le Bassin Artois-Picardie*, Rapport rédigé pour le Ministère de l'Environnement (SRETIE) et l'Agence de l'Eau Artois Picardie, L'OEIL/IUP, Université Paris XII, Créteil, septembre 1988, polygr., 100 p + annexes.

Larrue, C. (1988b), *La Pollution des Eaux d'Origine Agricole: Comportements Agricoles et Politique Publique. Deux Études de cas dans le Bassin Seine-Normandie*, Rapport Rédigé pour le Ministère de L'Environnement (DPP/Service de L'Eau), L'OEIL/IUP, Université Paris XII, Créteil, septembre 1988, polygr., 93 p + annexes.

Larrue, C. (1991) *Impacts et Conditions de la Mise en Oeuvre de la Lutte Contre la Pollution d'Origine Agricole; Etudes de cas en Région d'Élevage Intensif et en Région Viticole*, L'OEIL/IUP Université Paris XII, Créteil, août 1991, polygr., 121 p.

Larrue, C. and Navarre, F. (1990) *L'efficacité Différentielle des Réseaux d'Eau et des Réseaux d'Assainissement: un Essai d'Explication*, Projet de recherche pour le Plan Urbain, OEIL/IUP, Université Paris XII, 9 p.

Maintz, R. (1989) Les bureaucraties publiques et la mise en oeuvre des politiques, *Revue International des Sciences Sociales*. **31**(4), 677–690.

Musgrave, R.A. (1959) *The Theory of Public Finance*, McGraw-Hill, New York.

OECD (1986) *OECD and the Environment*, Paris.

OECD (1989a) *Economic Instruments for Environmental Protection*, OECD, Paris.

OECD (1989b) *Water Resource Management, Integrated Policies*, OECD, Paris.

OECD (1991) *Environmental Policy, How to Apply Economic Instruments*, OECD, Paris.

Padioleau, J.G. (1982) *L'Etat au concret*, PUF, Paris, 222 p.

Papon, F. (1991) *Les "Routes de Première Classe": Une Tarification Différenciée* de *la Circulation en Agglomération pour en Améliorer l'Éfficacité Économique de Manière Socialement Équitable*, Thèse de troisième cycle, INRETS/Université Paris XII, 4 tomes.

Prud'homme, R. (1981) *Le ménagement de la Nature*, Dunod, 212 p.

Prud'homme, R. (1984) Rôle des gouvernments centraux et des collectivités locales dans la gestion de l'environnement *Changing Roles for Local and Regional Government in Environmental Management*, WNU Science Press, Utrecht, pp. 53–68.

Weber, L. (1988) *L'Etat, Acteur Économique, Analyse Économique du Rôle de l'Etat*, Economica.

4. Arguments, Assumptions and the Choice of Policy Instruments

The case of the debate on the CO_2/energy tax in the European Community

ANGELA LIBERATORE*
European Policy Unit, European University Institute, S. Domenico di Fiesole, Italy

1. Introduction

An analysis of the arguments and rationales given by the actors in the field of environmental policy may help us understand why certain policy instruments are or are not adopted, and why they are or are not effective in dealing with the problems at hand. Rather than being a study on the implementation process, the following pages will focus on the design and selection of policy instruments. The reason for this is that problems inherent in the formulation of instruments will influence their adoption and implementation. Research on the implementation of policy instruments is necessary to discern the difficulties to be faced when theories are put into practice and general concepts are put into a specific context; but more attention to the very development of policy instruments is needed to understand why certain instruments (among the many possible ones) are adopted and are successfully or unsuccessfully implemented.

The hypothesis put forward here is that the features of arguments (such as their consistency and the realism of the assumptions which are made in the process of formulation of policy instruments), and the way arguments are utilized by the various actors will play a crucial role in the choice of instruments and influence their effectiveness. This does not mean that arguments are regarded as substitutes of power relations; rather they are powerful resources in the policy process.

An analysis of some of the main kinds of argument and their corresponding assumptions (no pretence is made to provide an exhaustive typology) is provided in the first part of the paper. Then a discussion of the role of arguments in the policy process is offered. Finally, the case of the debate on the CO_2/energy tax in the European Community is examined to "test" the above mentioned hypothesis in a specific context, and some concluding remarks are made.

Bruno Dente (ed), Environmental Policy in Search of New Instruments, 55–71.

2. The Arguments and Their Assumptions

When advocating or opposing the adoption of certain policy instruments various kinds of arguments can be (and have been) made:

1. *political arguments*, which focus on the need (for politicians) to take, postpone or avoid action and to choose one or the other instrument depending on elements like institutional and economic feasibility, social acceptability, ranking of issues in the political agenda, and so on;
2. *economic arguments*, which point to the cost-effectiveness of the available options/instruments and, as far as environmental problems are concerned, to the need to internalize in the market system environmental externalities;
3. *legal arguments*, which stress the need to enforce existing legislation or to introduce new laws in order to oblige individual and collective actors to restrain from activities that damage the environment;
4. *scientific arguments*, which support (to a degree depending on the scientific uncertainties and controversies involved) the adoption of certain policy instruments by providing data and explanations about the causes, scale and timing of environmental problems and processes;
5. *ethical arguments*, which emphasize normative issues such as the right of present and future generations to a sound environment and the importance of fairness in dealing with the distributive aspects involved in the adoption of the various policy instruments.

These arguments are, in turn, based on several assumptions concerning nature, technological development, the behaviour of individuals and collectivities, the way institutions work, and the characteristics of the context in which the instruments should be applied.

Behavioural assumptions[1] are made with respect to, for instance, whether individuals are utility maximizers, whether they are committed to certain values, whether they obey laws because they feel this is right or only because they are afraid of sanctions, whether they are able to learn, and whether the behaviour of a collective actor can be regarded as the sum of the behaviours of its individual members.

To illustrate some of the problems that may arise with respect to behavioural assumptions, let us take the case of the assumption that individuals are utility maximizers. While it is perfectly reasonable to assume that entrepreneurs are utility maximizers, it can be misleading to assume that other important economic agents, such as consumers, are also driven by utility maximization calculations in choosing goods and services or in utilizing energy.[2] Moreover, while utility maximization behaviours can be observed in areas outside the economic realm (for instance, during electoral competitions), not all individual and collective behaviours can be explained in terms of utility maximization.[3] Of course, problems will arise if the assumptions that individuals obey laws or act on the basis of altruistic values are unduly generalized.

So, if not enough attention is paid to the realism and implications of behavioural assumptions, unreliable theories and models will be developed – even when credible arguments can be found to support them – and unreliable policy instruments will be designed. This is especially the case with those policy instruments aimed at changing the behaviour of target groups, but also with those aimed at changing environmental elements or processes since the introduction, for instance, of air filters or water treatment plants implies that once introduced they are well maintained and utilized.[4]

As far as *assumptions about institutions* are concerned, they are the result of analysis and evaluation regarding whether the institutions relevant for the problems at hand (local and national administrations, legal bodies, national and international financial organizations, etc.) have the resources and capabilities needed to perform their task, whether they are adaptive to changes in the socio-economic environment and whether it is possible to reform them and/or establish new ones.

Problems may arise also with respect to this sort of assumption. If it is assumed, for example, that local administrations have enough money and specialized personnel to ensure that ambient or emission standards are met and this is not the case, lack of or poor implementation of certain regulations will result. Similarly, if it is assumed that the existing taxation systems work well or can be easily reformed and this is not true, the adoption of fiscal incentives may be problematic.

In other words, if no explicit and careful consideration is paid to the actual working of institutions (including intra- and inter-organizational bargainings, institutional inertia, etc.), the formulation and implementation of any kind of policy instrument will be based on rather unsteady ground.

Assumptions about technological development are another important element in the design and adoption of policy instruments.

Advocates of "technology forcing" regulations explicitly or implicitly assume that cleaner and/or more energy efficient technologies are only developed and diffused when industries are obliged to meet certain standards. If it is instead assumed that an environmental-friendly technology development is an intrinsic component of industrial processes (for instance, due to international competition), incentives for supporting such development will be argued for. Moreover, more or less optimistic/pessimistic assumptions can be made regarding the speed of the desired technological progress and about the ability to manage technologies in an effective and socially acceptable way.

To take for granted one or the other assumption may lead to the adoption of policy instruments that then prove to be ineffective.

Assumptions about nature play a crucial role in the choice of instruments for environmental policy.

Assumptions about the limited or unlimited "carrying capacity" of ecosystems, about their limited or unlimited ability to absorb pollution and/or restore themselves involve very different consequences with respect to the choice of policy instruments. For instance, those who think that nature is "ephemeral" will favour the ban of certain activities or products together with strict and binding regulations of others, while those who think that nature is "benign" will adopt a *laissez-faire* attitude.[5] Moreover, those who regard nature protection as a value in itself will tend to oppose any attempt to attach a monetary value to natural resources, while those who emphasize the need to address the economic causes of environmental disruption by internalizing environmental externalities will favour the adoption of economic instruments.[6]

Also with respect to the assumptions about nature, an insufficient consideration of their reliability may bring about the choice of ineffective or even damaging actions and instruments.

Assumptions about the context involve the above mentioned assumptions plus assumptions regarding the acceptability of the various policy instruments by different social groups and/or different countries, about the relative stability or instability of political systems and economies, and other elements. Wrong assumptions about the context (local, national and international; social, economic, political) where policy instruments should be applied may involve undesirable consequences.

A mix of these kinds of assumptions underlies the formulation and the choice of policy instruments.

Arguments suggesting the adoption of **incentive instruments** are based on the assumption that individuals as well as organizations are utility maximizers, that administrations have not enough resources and are not flexible enough to effectively implement direct regulations, that natural resources can be given an economic value, that ecosystems can tolerate certain forms and quantities of pollution, that clean technologies will be more quickly available through economic incentives, that these incentives are socially acceptable and that macroeconomic consequences will be positive or moderately negative.

Authority/command-and-control instruments are argued for on the basis of the assumption that individuals and target organizations are willing or can be obliged to obey laws, that political and administrative institutions have the necessary resources (technical, economic, legal resources but also authority and legitimation) to enforce regulations, that clean technologies will be more quickly developed if industries are forced to produce and use them, that certain forms and quantities of pollution must be absolutely avoided because they are unsustainable for the natural environment, that direct regulations are socially acceptable and that they do not represent too heavy a burden – or at least that they represent only a necessary burden – for the economy.

Arguments for introducing **learning and capacity instruments** assume that individuals (independent of their being utility maximizers or committed to altruism and intangible values) are able to understand and use information – if made available – to draw lessons from experience, and to modify their perceptions and behaviour accordingly. These arguments also assume that institutions are both able and willing to provide an arena for information and debate, that cleaner and/or energy efficient technologies will be more diffused it their benefits are made known and people are trained to use them. In case only the introduction of learning and capacity instruments is advocated (which is almost never the case) it must also be assumed that nature is "forgiving and patient" enough to wait for long-term cultural and behavioural changes to occur.

It must be pointed out that a "black and white" approach to the choice of policy instruments (or command-and-control tools or incentives, each of them eventually supplemented with learning and capacity instruments) is rarely argued for. Rather various types of "mix" are usually suggested such as the combination of taxes, product standards, promotion of research and other measures suggested to achieve the CO_2 stabilization target determined at the European Community level.[7]

"Mixes" can be argued for by saying, for instance, that certain kinds or quantities of pollution must be absolutely avoided because the ecosystem cannot tolerate them, but that the best (cost-effective) way to achieve this is to rely on incentives aimed at changing entrepreneurs' and/or consumers' behaviour.

Let us examine now the process which brings about the choice of certain policy instruments and mix of instruments.

3. The Role of Arguments in the Choice of Policy Instruments

The choice of policy instruments is part of a complex process where problems have first to be identified and the need for action acknowledged in order to start thinking about which tools should be selected.

However once this stage is reached, the various parts of the policy process influence each other rather than follow an orderly sequence. For instance, the availability and selection of certain instruments may determine the goals to be achieved (rather than the contrary), and the problem initially framed in a certain way may be re-framed at a later date due to new information, experience and evaluation of what has been done in the meantime.

The policy process as a whole is characterized by communication between the various actors involved – communication which is asymmetrical (since the actors have not the same economic, political and technical resources), and which shapes the framing of issues, the setting of goals, the formulation of strategies, the design and choice of instruments and their implementation. In the context of the communicative interaction[8] which shapes all these stages/components of the policy process, the features of arguments and the way they are utilized (in a more or less persuasive manner, with support or lack of scientific evidence, etc.) play a crucial role.[9]

This does not mean that sound arguments can substitute for existing power relations: good arguments are not enough to ensure that less powerful players are listened to and have weight. However, arguments can be a powerful resource; at least if coercion or manipulation do not prevent them from being made and communicated. If used in a way to influence the perceptions and 'capture' the interests of – and build coalitions with – other actors, arguments may in fact help those players with less economic or political resources to succeed in modifying the terms of reference and/or the ranking in the policy agenda of the problems at hand.

When focusing on the choice of policy instruments, the interaction between policy makers and experts (such as economists, lawyers and policy analysts) is especially important as far as the formulation of options (single instruments and possible mix of instruments) is concerned. Depending on various factors (like the prevailing of pluralist or cooperative settings, the relative transparency or secretiveness of the decisional processes, etc.) public and private interest groups may exercise a decisive influence in designing and even more in selecting one or the other possible option.

Different actors usually refer, at least in the beginning, to different arguments when advocating or opposing the choice of policy instruments. The differing positions and arguments can be sketched as follows.

Most economists, but also an increasing number of policy makers and interest groups, argue for repairing market failures through market mechanisms and for selecting the most cost-effective options. Lawyers, and all those who do not 'trust' market mechanisms, emphasize the need for norms in order to protect health and environment. Policy analysts and administrators usually point to institutional resources and constraints and to the importance of taking them into account in choosing among instruments. Policy makers tend to favour politically 'safer' options and to choose one or the other policy instrument depending on several elements, including previous experience and pressures from public and private interest groups. As far as these last actors are concerned, they will advocate or oppose certain policy instruments on the basis of their interests, values and assumptions about nature, the working of institutions, and so on.

In the debate that accompanies the bargaining and negotiations over the selection of appropriate policy instruments, such arguments are considered. Once the debate and negotiations have advanced enough, a process of learning develops in which new, as well as previously disregarded, knowledge and information are diffused, a common terminology is developed, self-reflection on previous experience (including previous policies) is favoured, attention is paid to aspects previously neglected and issues are re-framed accordingly.

The development of such a learning process does not necessarily imply that the 'best' option will be chosen and that improvements in the management of the problems at hand will be made. It only implies that options are evaluated taking into account the available information, points of view and experience.[10]

In such a process various factors can make an argument, or combination of arguments, prevail and ensure certain options are chosen. These elements are mainly the power or prestige of those who put forward the arguments, the inherent strength of the various arguments (their consistency, their forming part of a coherent 'paradigm', the realism of the assumptions implied), and especially the way they are made and utilized. For instance, by choosing the 'right time', by providing evidence, by forming coalitions thanks to the ability to convince the others that the available evidence supports a certain argument, that certain aspects that look negative are eventually not so, and so on.

In the next section we will analyze how this happens in a specific context.

4. The Debate on the CO_2/Energy Tax in the European Community[11]

To understand why and how the CO_2/energy tax had been selected as the main policy instrument to tackle the greenhouse issue at the European Community level, the evolution of the issue will be examined through a brief 'chronology' and the arguments elaborated and used by the actors involved will be investigated.

The analysis focuses on the debate within the E.C. Commission since this is the institution where the tax proposal had been elaborated. Inputs from other E.C. institutions (mainly the E.C. Council and the European Parliament), from member states, from environmental NGOs and business organizations will be only marginally referred to. The reason for this is that while these inputs (especially from the Council) had been very important in ensuring some specific aspects were emphasized or neglected, the Commission remains the 'place' where they were selected, further developed/reframed and utilized to modify – and try to make accepted – (its own) previously initiated proposals.

4.1. *Chronology*

The risk of global warming resulting from anthropogenic activities (i.e., emissions of greenhouse gases in sectors like energy production, transport, agriculture, etc.)

was firstly addressed at the E.C. level as a research issue and was introduced in the 1979 research programme on climatology (published in Official Journal of the E.C., O.J. L 12, 17 January 1980).

After being confined to E.C. research activities for several years, the risk of global warming was firstly addressed as a policy – beside scientific – issue by the European Parliament in July 1986. Two reports submitted to the Parliament in that month, specifically those submitted by Undine-Uta Bloch von Blottnitz and by James Fitzsimons, addressed the greenhouse issue.

The greenhouse issue is also mentioned in the Fourth Environmental Action Programme submitted by the Commission – Directorate General on Environment, Civil Protection and Nuclear Safety (DG XI) – to the Council of Ministers in October 1986. In that programme it is stated that: 'In case further research should confirm the likelihood of such impacts (i.e., impacts of the greenhouse effect on climate and agriculture), the Community should already be thinking about possible responses and alternative energy strategies' (O.J. C 70, 17 March 1987, p. 13). Interestingly enough, the reference to possible responses and alternative energy strategies is not included in the final text of the Fourth Environmental Action Programme (O.J. C 328, 7 December 1987) where the greenhouse effect is only referred to as a topic of the E.C. research programme on climatology.

A few months later, in July 1988, an interservice group (that is, a group formed by officials of various DGs) was established within the E.C. Commission to examine the problems related to the greenhouse effects. The creation of such a group indicates that the Commission as a whole had decided (mainly due to external events such as the Toronto Intergovernmental Conference called "The Changing Atmosphere. Implications for Global Security", held in June 1989) to devote more political attention to those problems.

In November the same year the Commission approved the draft "Communication to the Council on the Greenhouse Effect and the Community" submitted by the interservice group. The Communication (COM (88) 656 final) is devoted mainly to a summary of the state of knowledge based on the results of international scientific conferences. As far as response options are concerned, the document mentions various measures related to the energy sector, forests, agriculture, seas, CFC emissions. Since the adoption of that Communication the greenhouse issue is continuously present in the Commission's agenda.

In June 1989 a Council Resolution on the Greenhouse Effect and the Community (O.J.C. 183, 20 July 1989) was adopted where it was emphasized that the Community must make an effective contribution to the wider international debate. In subsequent documents of the Commission produced by the end of 1989 (COM (89) 369 final) and in the following years, a shift in thinking appears, from "making an effective contribution" to "take a leading role" in the international debate and negotiation on the greenhouse issue.

In the meantime (starting from May/June 1989) the Commissioner for Environment, Carlo Ripa di Meana, announced and reiterated the idea of introducing a tax on carbon dioxide (see Agence Europe, 8–9 May 1989, 29–30 May 1989, 8 June 1989, 6 December 1989).

In 1990 the main elements of the E.C. greenhouse policy were more precisely defined and elaborated.

On the occasion of international events dedicated to the greenhouse issue such as the White House Conference held in April 1990 and the Bergen Conference held in May 1990, the Commissioner for Environment attacked the U.S. wait-and-see position and strongly emphasized the idea of E.C. environmental leadership. Such an idea was also taken up during the joint Energy/Environment Council held in October 1990 (see Minutes of the Council 29 October 1990) and in a Commission's Communication dated 28 November 1990.

During the mentioned Energy/Environment Council, just before the Second World Climate Conference and mainly in order to establish a Community position for that occasion, the overall goal of E.C. greenhouse policy was set by adopting the Decision to stabilize the total CO_2 emissions by 2000 at the level of 1990 in the Community as a whole. In the foreword of the Council Decision (and in various Commission's document as well) reference is made to the findings of the Intergovernmental Panel on Climate Change (IPCC, 1990) which are regarded by the E.C. institutions as the consensual and authoritative scientific basis/evidence to guide action.

In the meantime the Directorate General on Energy (DG XVII) produced a study – based on a previous study by DG XII (Guilmot *et al.*, 1986) – and organized an international conference in Brussels on "Energy for the New Century: The European Perspective" (see the special issue of Energy in Europe, July, 1990). The event reinforced, or simply made more explicit, the energy-focused orientation of the E.C. greenhouse policy. While such orientation was mainly shaped by the Commission, it was also shared by the European Parliament who organized a public hearing on Energy and the Environment in September 1990.

In various documents, and in particular in a Communication to the Council jointly prepared by the Commissioner for Environment and the Commissioner for Energy (Antonio Cardoso e Cunha) in November 1990 ("A Community action programme to limit E.C. carbon dioxide emissions and to improve the security of energy supply", 28 November 1990), a Communication which was not adopted by the Commission, environmental and energy goals are linked and the overall E.C. strategy, the so-called "no-regret strategy", is emphasized. The main idea of the no-regret strategy is that, beside being beneficial for the environment, the actions suggested – mainly focused on energy efficiency aspects – are also beneficial for the economy.

Also during 1990, the use of economic instruments – and in particular the possibility/need to introduce a CO_2 tax – to deal with the greenhouse issue was advocated again by Ripa di Meana and also in the 'Report of the Working Group of Experts from the Member States on the Use of Economic and Fiscal instruments in E.C. Environmental Policy' (report XI/185/90) and other internal documents prepared by DG XI staff. Moreover, the idea of a combined CO_2/energy tax was mentioned for the first time in December 1990 in a working paper drafted by DG II (Directorate General on Economic Affairs) of the E.C. Commission on 'Policy Options in View of the Community's CO_2 stabilization target'.

The year 1991 was characterized by the debate over the overall policy package to be elaborated to meet the stabilization target decided in October 1990, and especially on the specific features (design) of the CO_2, then CO_2/energy, tax.

The main points addressed in the formulation of such a policy instrument had been the following: CO_2 vs. CO_2/energy tax, the proportional division of the CO_2 and the energy component of the tax, level of the tax, timing and graduality in the introduction of the tax, exemptions, neutrality vs. non-neutrality of the tax. Moreover the burden sharing issue and the integration of the tax with other policy instruments like research and development actions are also addressed.[12] The points mentioned can be found in various Commissions' documents elaborated in 1991 (see, SEC (91) 1744 final, document II/ 445/91-EN prepared by staff of DG II and other documents).

In the same year the Economic Policy Committee (EPC) discussed the issue of CO_2 emissions' limitation twice on the basis of documents prepared by DG II (II/82/91 and II/445/91). The chairman of the committee then sent a statement to the Commission and the Council summarizing the EPC's view on the proposed Community strategy. In the statement it was stressed that market-based instruments should constitute a major component of any comprehensive policy aimed at limiting greenhouse gases.

In May 1992, under the pressure of and just before another external event – i.e., the UN Conference on Environment and Development held in Rio in June (UNCED) – the Commission proposal for a CO_2/energy tax was presented at the Parliament and was discussed by the Council. Beside the already mentioned features, a very important element was added that eventually prevented any "unilateral" E.C. decision on a tax from being made. According to the "conditionality" introduced in 1992, the application of the tax will be in fact dependent on similar taxes or measures having similar financial effects being applied in other O.E.C.D. countries (Agence Europe, 28 May 1992, p. 10).

Since the arguments made and used in the formulation and in view of the adoption (or non-adoption) of the tax represent my focus of analysis, they will be specifically dealt with in the following section.

4.2. *Arguments Regarding the CO_2/Energy Tax*

The desirability, possibility and need to introduce a CO_2/energy tax had been advocated on the basis of various kinds of arguments.

The features of arguments, the evidence provided to support them, the position/power of (and coalitions between) those who made the arguments, the time and occasions when arguments were introduced or repeated are some of the factors that explain why and how the CO_2/energy tax had been selected as a main instrument to achieve the E.C. stabilization target.

Scientific arguments and evidence had been very important in the initial stage, i.e., in the identification of a greenhouse problem/risk and of some possible courses of action. For example, the available scientific evidence (IPCC, 1990) which establishes that CO_2 emissions are the major contributors to the greenhouse effect, have a long lifetime and their sources are known. This scientific evidence was used at the E.C. level to select these emissions as the main target for action. The assumption that underlies this use of scientific evidence is that to address the main, and more easily identifiable, cause of the greenhouse problem is a sufficient or good step to cope with it.

Moreover scientific evidence was also put forward to argue for the need to take preventive action. In this respect, scientific arguments were utilized in conjunction with the reference to the precautionary principle[13] included in the mentioned Fourth Environmental Action Programme and in the Single European Act.[14] Without such conjunction scientific arguments could have been used differently and a position similar to the one developed by the U.S. government, i.e., reduce the existing scientific uncertainties before taking any action, could have prevailed also within the Community.[15] It is worth mentioning that, within the Community, the precautionary principle is not only referred to on the basis of ethical arguments; even if the issue of intergenerational equity is mentioned in some occasions, the E.C. is also guided by the cost-effectiveness economic argument (in this case based on the assumption that prevention is generally less costly than reparation).

Political and economic arguments then shaped the formulation of the overall E.C. greenhouse policy and the design of the CO_2/energy tax.

Political arguments mainly regarded the need/desirability for the Community to take a leading role in the international debate and negotiations on the greenhouse/climate issue (especially in the perspective of the UNCED of June 1992). There was a corresponding need to prove the Community's real commitment by setting an overall stabilization target and deciding measures for meeting it. This argument implies that in order to be a 'world leader' the Community must strengthen its internal integration; in other words, to achieve an 'external relations goal' (be it environmental or of other sorts) the E.C. must fulfil its very constitutive integration goal.

Also linked with the integration issue is another, usually implicit but very important, political argument for the introduction of a tax at the Community level. According to this argument such a tax would represent a fundamental precedent for increasing the power of Community institutions in the fiscal arena and for increasing the overall Community economic and political integration. However some services of the Commission, mainly the Directorate General on Indirect Taxation (DG XXI), while sharing the overall integration goal, seemed to regard as negative/risky the fact that an 'environmental tax' could represent such a precedent.

Also regarding political/policy arguments, the issue of institutional feasibility and the lack of experience/familiarity within the Community with instruments such as tradable permits played a crucial role in putting forward the idea of a tax. In this case the assumption is made that it is better, or simply easier, to modify existing mechanisms and institutions rather than introducing new ones.

Other political arguments influenced the shift from an exclusively CO_2 tax to a CO_2/energy tax. In fact a CO_2 tax would favour nuclear power energy, and this would be hardly acceptable not only to those public pressure groups that oppose nuclear power (as voiced, for instance, by the European Environmental Bureau and the Climate Action Network, umbrella organizations of European NGOs) but also to the oil and coal industry/lobbies. On the other hand, this mainly politically driven shift had been also advocated in some E.C. sponsored studies (see, Capros *et al.*, 1991) on the basis of economic arguments, that is the economic benefits that would result from a tax that promotes energy efficiency.

Economic (cost-effectiveness) arguments provided the background and then the legitimation of a policy instrument mainly decided at the political level.

They provided the background since the debate on the adoption of economic instruments for environmental policy, initiated at the O.E.C.D. level, had spread at the E.C. (Commission) level, especially as a result of the publication of the 'Report on the Environment and the Internal Market' in 1989 (Task Force, 1989). An influential argument made in the E.C. context is that the existing Community Directives and Regulations are poorly implemented and that it is therefore necessary to partially substitute and partially complement them with economic incentives.

What is frequently stressed in the economic arguments for introducing a CO_2/energy tax is that such a tax is different from existing taxes since it is mainly aimed at influencing behaviour rather than raising revenues. In other words, an 'environmental reframing' (based on various assumptions about individual behaviour, the working of institutions, etc.) of that specific economic instrument called 'tax' is put forward.

The position and influence of those who supported economic arguments has been crucial in ensuring the tax instrument was selected. The use of economic instruments in general and the tax in particular has been advocated by members of various

Directorate Generals of the Commission (DG XI, DG II, DG XII and later DG XVII) as well as by the prestigious *Cellule de Prospective* (regarded as the 'think tank' of the Commission's President Jacques Delors) and by the Commissioner for Environment. Also the European Parliament started paying attention to, and favouring, the use of economic and fiscal instruments for environmental policy and organized a hearing on this topic in June, 1990. At the Council level, the Dutch and the Italian Presidency also explicitly backed the use of economic instruments and the tax proposal.

The successful spread of economic arguments had been also favoured by the connected use of legal arguments, mainly the reference to article 130 R of the Single European Act which states that the potential benefits and costs of action or of lack of action must be taken into account in E.C. environmental policy.

Besides providing the necessary background for the selection of a tax as a main policy instruments of the E.C. greenhouse policy, economic arguments have been crucial in 'legitimizing' – providing a sound/convincing basis for – an instrument mainly decided at the political level. In this respect, economic arguments had been interwoven with and reinforced by *technical arguments* and evidence.

On the one hand, energy models and studies promoted by DG XII within the framework of the Joule programme and regarding the cost-effectiveness and the macro- as well as microeconomic implications of a CO_2/energy tax (see Capros *et al.*, 1991; Coherence, 1991), provided the technical arguments and evidence for making the case that there is an energy saving potential that could be better exploited by means of a tax and without major macroeconomic impacts. A thoughtful overview of the economics of greenhouse policy prepared by DG II also concluded that a tax (but a tradable-permits system, too) would allow the exploitation of the existing energy saving potential (see Mors, 1991). These studies, together with the ongoing collaboration within an interservice group between DG II, XI, XII, XVII and XXI, and studies made within DG XVII itself were instrumental in definitively drawing in "the tax side" the still uncertain/reluctant DG XVII, but not DG XXI which found itself in a minority position.[16]

Supplementary technical arguments/evidence were then provided by a study of DRI on the impact of E.C. measures (including the tax) to control CO_2 emissions on the European industry (DRI, 1991/1992). The study, commissioned by DG XI, was aimed at offering policy makers and industry organizations an overview of the package elaborated within the Commission and its implications. Such a study, produced outside the Commission, represents an independent/external support to the Commission's proposal.

Within the above mentioned studies and models, and in Commission's policy documents, the specific features of the tax, i.e., its level, timing, graduality and neutrality, percentage between CO_2 and energy component were addressed. Economic arguments converged (eventually by chance) with the mainly politically driven proposal

– especially advocated by the Commissioner for Environment – of a tax of $10 per barrel of oil equivalent. Higher amounts (for instance, $18) had been discarded mainly on the basis of the argument that a higher tax would be needed only if the tax would be the only instrument adopted to achieve the stabilization target. Since other measures had been also designed and approved (such as the SAVE and THERMIE programmes for improving energy saving and efficiency) the $10 has been generally regarded as sufficient.

As far as the neutrality of the tax is concerned, both economic and institutional/ political arguments have been suggested, mainly the need (for economic, administrative and political reasons) not to increase the overall tax burden within the E.C. member countries.

Also the gradual introduction of the tax has been argued for on the basis of 'mixed' arguments involving issues of economic and administrative feasibility. Exemptions in exchange of voluntary agreements were included in the E.C. policy package mainly for political reasons, that is in order to overcome industry opposition.

The percentage of the CO_2 and energy component (50% each, or a higher percentage either of the energy or the CO_2 component) has been also debated by utilizing 'mixed' economic (cost-effectiveness), technical (fuel substitution possibilities) and political (acceptability) arguments.

Finally, economic arguments proved crucial in adding the terms of 'conditionality' of the proposed tax. Such arguments pointed to a Community goal which is as important as integration, that is the preservation and improvement of E.C. economy competitiveness in face of competition from the U.S., Japan and other O.E.C.D. countries. This was especially 'pushed' by some member countries (the U.K., Spain and Greece) and by some Commission services (mainly DG XXI) that opposed the original tax proposal. Supported also by other countries and services, such arguments can be hardly disputed on a purely economic ground and are used to prevent the 'unilateral' adoption of the E.C. tax.

5. Concluding Remarks

The process through which the above mentioned arguments (and their assumptions) had been developed, modified, linked, mixed and diffused within and outside the E.C. Commission is a very complex one. It can be however summarized in its main components as a learning process where – through asymmetrical (due to power relations) communication, including the building of a common language characterized by the pervasiveness of words like leadership, prevention, cost-effectiveness, energy efficiency, etc. – a problem had been selected for scientific and then policy attention. Where an instrument for dealing with such problem had been designed on the basis of some previously agreed principles and previous policies, but also by providing new evidence, reflecting about the difficulties at hand, negotiating about controversial problems, forming coalitions, re-framing issues. For instance, as far as the last aspect is concerned, by re-framing environmental protection issues also in terms of economic benefits and, on the other hand, by 'greening' non-environmental objectives like political leadership and economic competitiveness.

The formulation of the CO_2/energy tax can thus be regarded as the result of a learning process where political arguments (mainly referring to the leadership and integration issues) played an especially important role. These arguments were supported/reinforced by various scientific and technical arguments and by economic arguments concerning the general usefulness of economic instruments for environmental policy.

With respect to the adoption of the tax, economic arguments pointing to the need to preserve the E.C. competitiveness in the world economy prevailed and resulted in the prevention of the unilateral introduction of such an instrument in the Community.

We are thus faced with an economic instrument being designed and selected mainly on the basis of political arguments and not being implemented due to economic arguments and interests. This seems rather counter-intuitive since it is usually assumed, and is frequently the case (see, Larrue, 1995), that economic instruments are not implemented as a result of political and administrative reasons. Such a counterintuitive finding reinforces the importance of the way the political arguments had been made and used (by certain influential actors, in certain favourable moments, in conjunction with certain other kinds of arguments) in order to explain why a rather complex and economically controversial (as proved by the conditionality terms introduced at a later stage) instrument as the CO_2/energy tax had been designed at the E.C. level.

As previously mentioned, the fact that a learning process has been taking place does not necessarily mean that the CO_2/energy tax is "the best of all possible instruments", nor that all actors learned the same "lessons" during the process. While all the relevant actors learned to acknowledge the existence of a "greenhouse problem" and the need to face it, very different attitudes remain about which action should be taken and how.

In other words, learning to address a problem does not necessarily mean learning how to manage it. We are then left with the far from reassuring question about how much time is available to find out how to cope with global environmental risks.

Notes

* Present address: European Commission, Directorate General for Science, Research and Development, Brussels, Belgium.

1 A comprehensive examination of the behavioural assumptions of policy instruments can be found in Schneider and Ingram (1990).

2 An analysis of this issue with specific reference to energy use is offered by Stern (1986).

3 On this point see the different approaches of authors like Downs (1957) and Riker-Ordeshook (1968) emphasizing utility maximizing behaviours in the political sphere and those pointing to social identity formation (Pizzorno, 1983) and intangible values (Barry, 1982).

4 On the distinction between policy instruments as "detectors" or "effectors" – of behaviour or environment – see Klok, this volume.

5 An interesting typology and discussion of "myths of nature" (ephemeral, benign, capricious, perverse/tolerant) is provided by Schwarz and Thompson (1990).

6 Arguments supporting these different perspectives can be respectively found, for example, in Kelman (1981) and in Pearce *et al.* (1989).

7 This case will be discussed in the last part of the article.

8 The reference is to the notion of "communicative action" developed by Habermas (1984).

9 On the role of argument, evidence and persuasion in the policy process see Majone (1989). See also Ash (1988), who focuses on the dichotomy between economic arguments and non-economic arguments, and Lindblom (1965) regarding the processes of mutual adjustment that occur also through arguments.

10 On the issue of learning see, for instance, Eder (1987), Haas (1990), Hall (1990), Miller (1986). A comprehensive review of the literature on learning is provided by Parson and Clark (1991).

11 This part is based on ongoing research within the framework of the project 'Social Learning in the Management of Global Environmental Risks' directed by W. Clark of Harvard University. The reconstruction of arguments is based on the analysis of E.C. documents and on very useful interviews with members of the E.C. Commission. The chronology and subsequent analysis cover the period of formulation of the tax proposal up to 1992. After that year the debate continued but no major changes in the proposal were introduced. Up to summer 1994 the tax had not been adopted.

12 While being very important components of the E.C. greenhouse policy, these aspects will not be treated in the following (CO_2/energy tax focused) analysis.

13 On the precautionary principle see Cameron and Wade-Gery, this volume.

14 The Single European Act (SEA), adopted in 1987, amends the 1957 Treaty establishing the European Economic Community and adds a special section on Environment. The principle that preventive action should be taken in the field of environmental protection is stated in art.130 R of the SEA. On this point and on E.C. environmental policy more in general see, for instance, Koppen (1988), Liberatore (1991).

15 It is worth pointing out that also the position of the Bush administration was framed in terms of 'no-regret policy', but in the more restrictive sense of not setting any binding goal – such as stabilization or reduction targets – and take corresponding actions that could eventually reveal to be unnecessarily costly. Implicit in this position is the assumption that waiting costs, or costs of adaptive actions, are lower than the costs of preventive actions.

16 It is worth pointing out that the focus on energy turned attention aside (at least temporarily) from other relevant sectors like transport and agriculture, and thus from the positions of other Directorates General which are however likely to play a role in a future stage of the E.C. greenhouse policy.

References

Ash, P. (1988) *Consumer Safety Regulation. Putting a Price on Life and Limb*, Oxford University Press, New York, Oxford.

Barry, B. (1982) Political participation as rational action, in B. Barry and R. Hardin (eds), *Rational Man and Irrational Society?*, Sage, Beverly Hills, London, New Dehly.

Cameron, J. and Wade-Gery, W. (1995) *Addressing Uncertainty: Law, Policy and the Precautionary principle*, this volume.

Capros, P., Karadeloglu, P., Manztzos, L. and Mentzas, G. (1991) *Impact of Energy and Carbon Tax on CO_2* Emissions, Report for DG XII of the E.C. Commission, Joule Programme.

Coherence, D. (1991) *Cost-Effectiveness Analysis of CO_2 Reduction Options*, Report for DG XII of the E.C. Commission, Joule Programme.

Downs, A. (1957) *An Economic Theory of Democracy*, New York.

DRI (1992) *Impact of a Package of E.C. Measures of Control CO_2 Emissions on European Industry*, Final Report (draft report completed in November 1991), Prepared for DG XI of the European Commission.

Eder, K. (1987) Learning and the evolution of social systems. An epigenetic perspective, in M. Schmid, and F. Wuketis (eds), *Evolutionary Theory in Social Science*, D. Reidel Publishing Company, Boston.

Energy in Europe (1990) Special Issue, *Energy for a New Century: The European Perspective*, E.C. Commission, DG XVII.

Guilmot, J.F., McGlue, D., Valette, P. and Waeterloos, C. (1986) *Energy 2000*, published by Cambridge University Press on behalf of the E.C. Commission, DG XIII, Cambridge.

Haas, E. (1990) *When Knowledge is Power. Three Models of Change in International Organizations*, Univ. of California Press, Berkeley, Los Angeles, Oxford.

Habermas, J. (1984) *The Theory of Communicative Action*, Bacon Press, Boston.

Hall, P. (1990) *Policy Paradigms, Social Learning and the State: The Case of Economic Policy-Making in Britain*, Estudio/Working Paper 1990/4, Instituto Juan March, Madrid.

IPCC (1990) Policymakers' Summary of the Report of Working Group 1 on scientific assessment.

Kelman, S. (1981) *What Price Incentives? Economists and the Environment*, Auburn House Publ. Company, Boston.

Klok, P.J. (1995) *A Classification of Instruments for Environmental Policy*, this volume.

Koppen, I. (1988) The European Community's Environmental Policy: From the Summit in Paris, 1972, to the Single European Act, 1987. EUI Working Paper, Florence.

Larrue, C. (1995) *Political (un)feasibility of environmental economic instruments*, this volume.

Liberatore, A. (1991) Problems of transnational policymaking: environmental policy in the European Community, *European Journal for Political Research* (19).

Lindblom, C. (1965) *The Intelligence of Democracy. Decision Making Through Mutual Adjustment*, The Free Press, New York.

Majone, G. (1989) *Evidence, Argument and Persuasion in the Policy Process*, Yale University Press, New Haven and London.

Miller, M. (1986) *Kollektive Lernprozesse*, Surkamp, Frankfurt/Main.

Mors, M. (1991) *The Economic of Policies to Stabilize or Reduce Greenhouse Gas Emissions: The Case of CO_2*, internal paper, E.C. Commission, DG II.

Parson, E. and Clark, W. (1991) Learning to Management Global Environmental Change: A Review of Relevant Literature, Contribution n.8 to the project "Social Learning in the Management of Global Environment Risks".

Pearce, D., Markandya, A. and Barbier, E. (1989) *Blueprint for a Green Economy*, Earthscan Publications Ltd, London.

Pizzorno, A. (1983) Sulla razionalità della scelta democratica, *Stato e Mercato* (7).

Riker, W. and Ordeshook, P. (1968) A theory of the calculus of voting, *American Political Science Review* (62).

Schneider, A. and Ingram, H. (1990) Behavioral assumptions of policy tools, *Journal of Politics* (2).

Schwarz, M. and Thompson, M. (1990), *Divided We Stand. Redefining Politics, Technology and Social Choice*, Harvester Wheatsheaf, New York, London.

Stern, P. (1986) Blind spots in policy analysis. What economics doesn't say about energy use, *Journal of Policy Analysis and Management* (2).

Task Force (1898) *Report on the Environment and the Internal Market*, Brussels, E.C. Commission, DG XI.

5. Courts and Citizen Action

NICOLA PAIN
BALLM University of Sydney, Sydney, NSW, Australia

1. The Expansion of Environmental Law and Policy

The failure of the common law to evolve fast enough to encompass community concern for environmental protection necessitated that parliaments act to provide environmental protection legislation. Depending on the constitutional arrangements in certain common law countries, laws developed at both the national and state levels of government. In some federal systems of government, such as in Australia, the national government's powers in relation to the environment are limited. Hence the state and local governments have a major role to play.

In common law jurisdictions, such as the U.S.A., Canada, Australia, New Zealand and the U.K. environmental law has emerged as a dynamic field of law in the last two decades particularly, in the 1980s and now the 1990s. The U.S.A. has probably been the most active common law country in the development of environmental laws at both federal and state levels. The passage of major federal legislation such as the Clean Air Act and Clean Water Act in the 1960s was followed by complex legislation such as the Resource Conservation and Recovery Act 1976 (RCRA) and the Comprehensive Environmental Response, Compensation and Liability Act 1980 (CERCLA) (Superfund legislation). Canada and Australia have also developed their own environmental law substantially in the last ten to fifteen years. The U.K. has been slower in adopting comprehensive environmental laws but its recent Environment Protection Act 1990 (U.K.) is a major advance.

An important question is the way in which "environment" is defined in a legal system. Environment can be defined in ecocentric or anthropocentric terms in legislation. This influences how court action to protect the environment by citizens is regarded in the court.[1] Recognition of the need for ecosystems to be represented in the legal system is a major philosophical step for systems which traditionally focus exclusively on the way in which humans are affected.

There are now many kinds of environmental law. Categories include

- environmental protection laws, such as pollution control laws, environmental impact assessment laws and cultural heritage protection laws;

Bruno Dente (ed), Environmental Policy in Search of New Instruments, 73–93.

- planning and development laws, which regulate land use and, in Australia, are concerned with regulating the decisions made by local councils as these elected bodies are responsible for day-to-day land use decisions;
- nature conservation laws, which establish protected natural areas and aim to protect habitat, biodiversity, flora and fauna;
- resource laws which regulate the exploitation and use of natural resources.[2]

To match the expanding scope of environmental law there are now many bodies concerned with the administration and development of environmental law and policy. Government departments and regulatory agencies, private industry, community groups and individuals, professional bodies including environmental law associations and academic institutions all have a role to play. Many groups now lobby governments to seek changes in the development of environmental law and policy or changes in the application of that policy.

One final, preliminary point must be made. The methodological focus of this paper reflects the formal, centralist paradigm of law. Formal law, by which is meant that law made by parliaments and judges, is given high priority in attempts to solve environmental problems. The premise is that answers can be found within the legal system. However, formal law does not always achieve the end for which it was created. Better environmental outcomes may often be achievable through 'green light' and pluralist approaches.[3]

This paper will focus on one aspect of the development of environmental law and policy, being those aspects concerned with citizen participation and the legal enforcement of environmental law. Central to this issue is the role of courts.

2. Defining the Public Interest

Public interest issues are dealt with variously in different jurisdictions. In many common law systems based on British traditions, the Attorney-General, the principal law officer of the state, represents the "public interest". For numerous reasons, however, the executive arm of government does not always ensure proper enforcement of environmental protection laws. In Australia, where the Attorney-General is an elected officer whose department advises the government departments which may be the respondents in public interest cases political considerations can arise in the decision on whether suit should be brought.[4] Private action to enforce public interest protection laws is thus imperative.[5]

In the environmental law context citizens must "represent" the environment in public fora and when lobbying government and industry decision makers. Many groups will claim to be acting in the public interest when they do represent the environment to argue for its protection. The concept of public interest is arguably difficult to define and there has been active debate in NSW, Australia, on how the term is used. The concept often raises fierce opposition from industry, and those government departments which represent their interests, because they argue economic development is as much in the public interest as protection of the environment. In their view both activities can be said to be "public interest".

Another aspect of the interrelationship between courts and citizen action in the field of environmental law is the rise of public interest environmental litigation. This phrase is now used in most common law jurisdictions to describe litigation taken to protect the environment by citizens groups or individuals. Public interest litigation is contrasted with private actions where the principal aim of the proceedings is the remedying of harm suffered by one or only a few individuals. Most books written on environmental law in Australia, Canada and the U.S.A. refer to the phrase "public interest litigation" to mean environmental protection cases.[6] Characteristics of public interest environmental litigation include:

1. The relief sought does not benefit the applicant financially.
2. Issues raised are substantial, and may on judicial interpretation, clarify the law and add to precedent.[7]

In the environmental field it is increasingly accepted that citizens and interest groups have a role to play in representing the public interest. There is general discussion and acknowledgment of the importance of public interest actions by professional bodies (such as the National Environmental Law Association in Australia) and some governments.

The need to provide mechanisms for citizen enforcement in public interest litigation is not universally viewed as a good idea. Reliance on litigation does not provide a very satisfactory means of problem solving in society in general if this leads to increasing friction and break down in "neighbourly relations". In the environmental field, court action on issues which are often politically charged, as major developments can be, are not considered desirable by some governments. In NSW, Australia, the difference between the rights given to enforce planning and development laws and the pollution laws also demonstrate this. The planning and development laws have been enforceable by citizen since 1979. In the pollution control area, however, opportunity for enforcement by citizens has been very limited. This has changed only recently with the amendment of pollution legislation in December, 1991.

It is beyond the scope of this paper to detail every aspect of public interest environmental litigation for all relevant jurisdictions. As the jurisdiction in which to author practices in NSW, Australia, has one of the better developed environmental law systems in Australia, modelled in part on U.S.A. legal models that jurisdiction will be focused on. Australia is a federation of states which must interact with the rational government. Most environmental legislation occurs at the state level.

3. The Role of Citizen Action

In the environmental law field citizen action can play a significant role. Aspects of this role include:

1. involvement in formulating environmental policy and its implementation;
2. participation in the decision-making process for approvals of pollution control; investment and development applications to government;
3. standard setting of pollution licence levels in particular cases;
4. monitoring of developments both private and public to ensure compliance with environmental laws;
5. enforcement of environmental laws and regulations;

This paper will focus on two aspects, enforcement and monitoring.

Enforcement from a legal perspective involves citizens in court action in several jurisdictions. In Australia and the U.S.A. "public interest" litigation in the environmental field is widely regarded as a useful part of the environmental law system. Public interest environmental litigation is said to enable the private enforcement of public rights. This has been justified by several commentators and is eloquently summed up as follows:

> In the complex society in which we live it frequently occurs that citizens in great numbers and across a broad spectrum have interests in common. These, while of enormous significance to the society as a whole, do not involve the fortunes of a single individual to the extent necessary to encourage their private vindication in the courts. Although they are within the executive branch of the government offices and institutions . . . whose function it is to represent the general public in such matters and to ensure proper enforcement, for various reasons the burden of enforcement is not always adequately carried out by those offices and institutions, rendering some sort of private action imperative.[8]

Whose Responsibility Is It?

Responsibility for enforcement of environmental laws (particularly pollution control laws) usually lies with a government agency.[9] The agency can enforce pollution control laws in various ways. These can range from

(i) a negotiation approach with industry and other government departments encouraging a certain result without recourse to litigation;
(ii) internal agency review of proceedings where breaches of the law are suspected;
(iii) court proceedings in public seeking criminal and/or civil sanctions against offenders.

The same or other agencies will be responsible for land use planning, particularly in those jurisdictions where land use zoning is used to impose restrictions on land use. In Australia, the enforcement of planning and land use control laws can lie with a state agency and/or local council.

Giving discretion to government agencies about how law is to be enforced can be problematic, particularly if the agency is or is perceived to be industry "captured" in the case of a pollution control agency. Citizen enforcement actions can be a back up to government agency action (and inaction). In this sense litigation to enforce environmental laws is like the privatisation of environmental law enforcement.

What Policies are Needed for Citizen Action?

Specific legislative and administrative measures which encourage citizen participation and often action at the national and constituent state level include:

(a) public ***participation*** in the planning and development process and environmental impact assessment procedures,
(b) access to ***information*** under Freedom of information and Community Right-to-Know legislation and
(c) broad ***standing*** rights which enable anybody to take an issue to court regardless of whether they are affected by a particular decision or circumstances. These measures should be backed up by systems which provide resources to enable public interest litigants the opportunity to participate in litigation.

Public Participation

Public participation in environmental impact assessment procedures can be specifically provided for and regulated by statue. Failure to comply with statutory requirements for citizen participation can be remedied in courts by citizens if appropriate legislative and procedural mechanisms are provided.[10]

Access to Information

Access to information is a key aspect of providing means to citizens to enforce their rights. Giving access to information to citizens about government decisions is an important right backed up by administrative laws in some countries. This is an important adjunct to judicial review proceedings.[11]

Open decision-making procedures and adequate public participation provisions are just as important. These can be mandated by statute but the extent of government agency willingness to engage in this can be a reflection of the agency's attitudes to outside community involvement.

Standing to Sue

Access to courts by citizens is an important right in the eyes of many common law lawyers. Civil law systems can produce different results and views on the usefulness of court proceedings as law enforcement measures often vary in these countries. One example is Germany where there are relatively limited opportunities for third party litigation. Rights under the Basic Law are considered to be individual rights relating to "private" rights. Public interest plaintiffs have limited access to courts. The general administrative law code prohibits litigation of claims that are not private rights (unless otherwise specifically changed in statute). None of the German environmental statutes authorise citizen rights.[12]

This is similar in practice to the English common law approach to standing. Under this doctrine, judicial review proceedings are concerned with individual rights.[13] Environmental values can be enforced through the courts if they also correlate to the infringement of a private right under strict common law standing rules.

If the court system is to be accessible to citizens it should mean that standing to sue in environmental protection cases is not limited to those plaintiffs with the appropriate "special interest", often financial. Accepting that the environment ought to have ecocentric values rather than human defined values should also mean that the environment is able to be represented without having the human plaintiff demonstrate a special interest in the subject matter or particular harm suffered in order to justify his or her standing to sue.

The provision of third party rights in courts is widely regarded as important for the overall operation of the environmental protection laws in NSW Australia. It is now accepted, even by conservative governments who have opposed citizen enforcement in the past, that these actions play an important role in ensuring that governments and private developers comply with environmental laws.

On one view it is citizen enforcement of environmental laws which is the final guarantee of enforcement.[14] This arises from the experience of government agencies which can become "captured" by constant contact with the industry they regulate, rendering them unable to fulfil this task adequately.

Australia provides interesting case law on the history of the development of standing rules in environmental cases. This is an on-going process as many state jurisdictions, as well as federal, still limit access to courts through narrow rules of standing or a narrow judicial interpretation of standing provisions in legislation.[15]

Some Environmental "Standing" Cases

The legislative provisions defining standing determine the context in which standing issues are made. The following case examples involve different pieces of legislation and are illustrative of the importance broad standing provisions in legislation can have for citizen action in Australia.

In 1980 the Australia Conservation Foundation (ACF) sought to challenge national law concerning foreign investment rules. The Foundation is the largest national conservation body in Australia. The Foundation was concerned about particular investment in an environmentally destructive tourist development in one state, Queensland. The main issue in the proceedings did not get to the court as the Foundation was held by the High Court of Australia[16] not to have sufficient interest to warrant it having standing regardless of the strength of its legal arguments. The position for challenging national laws may be in the process of changing as demonstrated by the next example. There is not clear authority to settle this issue as the High Court has yet to hear another case on this issue.

In 1989 the ACF challenged[17] a different national law[18] which regulated the issuing of export licences by the national government for woodchips to a Japanese company under mechanisms provided by administrative review legislation.[19] The legislation provided that "any person aggrieved" by a decision could take action to challenge its legality.[20] ACF was held to have standing by a single judge of the Australian Federal Court to take these proceedings because of its long involvement as an organisation in forest protection issues and as a nationally recognised conservation organisation. The decision suggests that some kind of "special interest" will still be required under the administrative review legislation as the local landowner who was the co-applicant in the proceedings was denied standing because his interests were not greater than any other member of the community.

At the state level, two cases demonstrate the different experiences applicants can have depending on the legislation they are taking action under and the level of acceptance public interest litigation has by the judiciary. In a case known as the Mount Etna Bat case[21] a group of speleologists and conservationists were trying to protect a limestone cave from destruction by a private quarrying company in the Australian state of Queensland. The cave was valuable habitat for a threatened species of bat. In hard fought proceedings under state fauna protection legislation the Supreme Court of Queensland held that the applicant did not have standing to sue in the proceedings under the relevant Queensland legislation. An appeal to the High Court was not proceeded with because of undertakings given to that court by the respondent that the issue would not be raised again in the State court.[22] Unfortunately on return to the Queensland Supreme Court the respondent was successful on another procedural issue. An application to the State court asking that the applicant provide further security for costs was successful. (The applicant had already provided a large amount of money in response to an earlier security for costs order.) As the applicant did not have the financial means to comply with the order its court action could not proceed. The cave in question was blown up in the couple of days after the court action had to be abandoned by the speleological society.

In litigation in NSW in 1989, a similar factual issue arose concerning the protection of a cave which was valuable habitat for a species of bat. At issue was the application of environmental impact assessment legislation, the Environmental Planning & Assessment Act (1979) NSW. This legislation provides that "any person may take action to remedy or restrain a breach of the legislation. The standing issue does not therefore arise as an issue. The legal case concerned the relationship between the Environmental Planning & Assessment Act 1979 (NSW) and the Mining Act 1973 (NSW). Ultimately the applicant, a member of a speleological society, was successful on appeal from the Land & Environment Court of NSW to the Supreme Court of NSW in holding that companies conducting mining activities were bound to comply with the state's main environmental impact assessment legislation as well as any requirements of the mining legislation.[23]

Statutes Must Overcome Limits on Standing

In the environmental field legislation has overcome the limits of the common law in areas such as standing to sue. The main piece of NSW environmental planning legislation, the Environmental Planning & Assessment Act 1979, provides broad standing for citizens to take action to enforce that law. Standing is provided for any person or group to restrain or remedy a breach of that Act. At the time the Act was drafted the provisions of the principal Michigan environmental statute, which paved the way in the U.S.A. for broad standing provisions in legislation, was adopted. Several other NSW environmental statues contain similar standing provisions in the conservation and heritage areas and the control of chemicals.[24]

Pollution Control

Pollution control and enforcement is an area where jurisdictions differ widely. This is in part because pollution law enforcement involves questions of criminal law. Pollution is controlled in many jurisdictions by means of a licensing system. Failure to comply with a licence or polluting without authorization, i.e., a licence, can given rise to an offence which is criminal. Unlike the U.S.A. where there has been civil enforcement of pollution laws for many years there has been reluctance by legislatures in Australia to open pollution laws to civil enforcement action. Such reluctance is probably a result of inexperience. "Penalty" implies criminal law, and the "civil penalty" concept is novel.[25]

Increased public awareness about environmental harm from pollution in Australia has led to much greater fines for polluters. In the 1970s and early 1980s fines which were issued were often very small, possibly reflecting the fact that the community itself did not consider that these offences were particularly significant. Legislation in NSW has now been introduced which provides for much greater penalties for pollution offences.[26] It also provides for gaol terms for gross offenders of up to 7 years and lifts the corporate veil on company directors.

The responsibility for prosecution of pollution offences has always been regarded as the exclusive prerogative of government agencies or officers both at state and local level. Standing to take criminal action has only ever been available for those persons duly authorised by the responsible Minister or the state government agency responsible. Under the Environmental Offences & Penalties Act (1989) NSW there is now wider scope for criminal prosecution by private citizens.[27] Given the much greater onus which the prosecution bears in these circumstances criminal enforcement actions by citizens is not likely to occur very often.

Enforcing pollution laws is an area where citizen enforcement by way of prosecution or by civil proceedings has not generally been encouraged in Australia, unlike the U.S.A. In NSW until 1990 there was no statutory mechanism for citizens to use to take third-party action in pollution enforcement cases.

The role of citizens in civil enforcement of pollution laws has now been broadened with recent changes in NSW. The Environmental Offences & Penalties Act 1980 (NSW) offers a critical stopping stone to better enforcement of pollution controls by civil means. Under 1991 amendments made to section 25 of that Act, any citizen can seek a restraining order from the Land & Environment Court to prevent the breach of any law causing or threatening to cause harm to the environment provided the court gives its consent.[28] This section removes the necessity to seek consent from the pollution control agency before civil action can be taken in the Land & Environment Court of NSW. It still places an obstacle in the path of those citizens wishing to take such action as it requires that the court itself must give leave to the taking of such actions. In order to make these orders the court must be satisfied that "there is a real or significant likelihood that the requirements for the making of and order under this section will be satisfied." This wording has not appeared in any other legislation of its type and its interpretation is being determined in proceedings against the NSW Environment Protection Authority commenced in the Land & Environment Court in May 1992.[29]

4. The Role of Courts

Courts can provide a vital role in the *enforcement* of environmental law. That role raises interesting questions about the relationship between courts and parliaments. This is particularly so in the case of environmental law because of the nature of the issues raised.

Civil Versus Criminal

Distinction should be made here between civil and criminal proceedings as these are obviously quite different in scope and purpose. Criminal prosecutions have not traditionally been the province of citizen enforcement actions in the environmental law field in Australia. Criminal law enforcement requires expertise and standards of proof which are very difficult for citizens to attain. Enforcement of the criminal law in general, and pollution and other environmental criminal laws in particular, is usually the responsibility of government regulatory agencies. Citizen action here is taken to be civil proceedings.

Power Sharing

Citizens may engage in political action and lobbying to try and have their view influence decision-making and policy. Courts can provide an additional power sharing mechanism between governments and citizens. Court action provides the means to otherwise legally powerless groups and individuals to take action to enforce rights and duties against private companies and governments. This can include action under environmental laws in civil proceedings. Court action rarely provides the solution to a given issue from the point of view of a citizens group, reflecting the relativity limited role courts traditionally play in reviewing government decision making. It is an important tool for citizens groups to use however.

In some jurisdictions citizens can take action forcing agencies to act. For example, citizens in the U.S.A. can petition for executive discretion to be exercised to make emergency response plans under the Emergency Relief and Community Right To Know Act in the event of or threat of a chemical disaster. Similarly, where legislation enables detailed environmental plans to be made, but government agencies do not do so, citizens can petition for action, and ultimately have their own plans endorsed as binding under the legislation. Of course, rigorous procedures must be complied with. Citizens can, however, overcome institutional paralysis in important site-specific issues such as the creation of management plans for environmentally sensitive areas.

The threat of court action alone can provide a powerful voice to citizen groups. Court action may not have to be proceeded with as the threat of this alone may cause the citizen group to be taken seriously enough for meaningful negotiations to occur with governments and private companies.

The Nature of Legal Proceedings

(i) *Judicial Review of Government Administrative Decisions*

The role of courts has also been developed to include, in many jurisdictions, the power to review government decisions. This judicial review function is an extremely important aspect enabling the court to review government decision making. Generally the court's role is restricted to reviewing the process by which government decisions are made. It cannot implement its own decision. Despite this there can be tension for governments in judicial review proceedings as the limited nature of these proceedings is not always clear to government. In the environmental field, judicial review proceedings can mean court scrutiny of circumstances where the government makes decisions which have adverse environmental impacts. Losing these kinds of actions can be embarrassing for governments. In NSW this has led to attempts to legislate away the effects of court decisions which have required the state government to comply with environment impact assessment procedures.

(ii) *Standing in the Decision-Maker's Shoes*

There is scope in the environmental field for a change in the role of the judiciary with the opportunity given to some tribunals and courts to decide planning questions on their "merits". In these cases the court's decision will stand in place of the decision-makers (for example in the Australia context this can mean a local council's planning decision). This is a role not often accorded to courts and can create controversy where a proposed development is significant economically yet has adverse environmental impacts. These actions also question public (government) decisions as well as private.

Specialist judicial officers may be needed to ensure that the necessary expertise exists in the court to judge subjective matters such as planning issues.

The Rise of Public Interest Environmental Litigation

An interesting and important debate in NSW and other Australia states has focused on the notion of public interest litigation. The concept of public interest environmental litigation is firmly entrenched in the legal system in Australia to mean actions to ensure protection of the environment. In the discussion below concerning matters in specialist environmental courts the argument concerning whether or not a matter is public interest can be an important issue in terms of finding funding for litigation and to avoid cost penalties and other procedural hurdles.

In Australia, court action is taken by incorporated resident action and conservation groups, by individuals on behalf of unincorporated groups of residents and individuals alone. The latter will generally have the support of a group of people but these will not be named in the court proceedings as a party.[30]

5. Specialist Environmental Courts

The role of courts is important in the enforcement of environmental laws, particularly in relation to government development. One important consideration is the role of specialist courts which exist in some jurisdictions to deal with environmental laws. A direct comparison of all the courts or tribunals in relevant jurisdictions is beyond the scope of this paper. One example will be considered in depth.

Land and Environment Court of NSW, Australia

A specialist environment court exists in NSW, Australia. The Land & Environment Court was established by statute in 1979.[31] It is equivalent to, but separate from, the Supreme Court of NSW, the highest court in that state . The Land & Environment Court deals exclusively with environmental, developmental and planning, pollution and heritage law in NSW.

The Land & Environment Court has two kinds of judicial officers, judges and assessors. The assessors are selected for their non-legal skills such as planning expertise as they must decide the merit appeal cases on individual development applications. As these involve subjective decisions about planning issues, their decisions can create unhappiness at the local council level. Assessor decisions on the merits of a particular development become the decision of local councils who must then implement them. Judges hear all cases involving legal issues, including civil enforcement actions and criminal prosecutions under the pollution laws.

Cases Before the Court

What types of cases are heard by the Land & Environment Court? The court has five classes of cases. Class 1 and 2 matters are concerned with development and building appeals. Dissatisfied applicants for development consent and building consent can appeal to the court. In these cases the court stands in the shoes of the local decision-making body, a democratically elected local council. Assessors in the court make a value judgement on the merits of the particular issue about whether or not the matter should be approved and can impose conditions of consent. Class 3 concerns land valuation matters. Class 4 is the civil enforcement class also described as the Court's equitable jurisdiction because the court can issue remedies such as declaratory orders and injunctions. Class 5 are pollution prosecutions generally taken by the NSW Environment Protection Authority.

In the civil enforcement proceedings, court action will generally be taken by an individual or group against a government agency or department and/or private company. Orders are sought from the court that certain decisions or actions are illegal. Injunctions are sought in appropriate cases. No individual damages are claimed in these kinds of actions.

NSW legislation provides for citizen participation in decision-making in relation to development proposals by both the government and the private sector. This includes participation by way of comment in the environmental impact assessment process. A breach of the mandatory provisions of an Act in relation to public consultation can give rise to legal proceedings seeking a declaration that a particular development approval is void because the provisions of an Act were breached.

Numerous cases have now been taken against government authorities by citizens in an individual capacity or in groups to enforce the environmental impact assessment requirements of the NSW legislation. Some examples of these cases are as follows:

(a) *Enforcing Environmental Impact Assessment by Government Agencies (Class 4 Proceedings as These are Judicial Review Proceedings of Agency Decision-making).* Wendy Jarasius was a resident of southern NSW who was, like many in her community, concerned about the logging of native forests. Logging operations were carried out under licence from the Forestry Commission of NSW which did not carry out any environmental impact statement prior to issuing these licences. Ms Jarasius commenced proceedings in the Land & Environment Court of NSW seeking declaratory orders that the Forestry Commission was in breach of the Environmental Planning & Assessment Act (NSW) for failing to do an EIS before issuing the licence. The test in deciding whether or not an EIS had to be done was whether the activity of logging would have a significant effect on the environment. The Commission argued that the activity did not. After six weeks of hearing the evidence of a large number of experts the court held that there was a significant impact and that therefore the Commission was in breach of the legislation and therefore acting illegally.[32] An injunction was issued which required that an EIS be done before any more logging could take place.

Other cases have been taken successfully against the Electricity Commission of NSW requiring it to undertake environmental impact assessment before deciding the siting of transmission lines.[33]

(b) *Reviewing Local Council Decisions over Private Development (Class 4 Proceedings as These Involve the Review of Local Council Decision-making).* Many actions have been taken by concerned resident action groups in the Land & Environment Court to challenge decisions of local councils in approving environmentally harmful developments. These actions are judicial review proceedings in which citizens argue that there are procedural irregularities in the way councils have made their decision, often because of inadequate environmental impact assessment as required under the Environmental Planning & Assessment Act (NSW). There are a large number of these kinds of cases where the court must review the decision-making procedure of a local council. In one case[34] the local council tried to avoid having to prepare an environmental impact statement for a waste management facility by incorrectly classifying the development so that this was not required. The successful appeal to the Land & Environment Court held that it was not reasonably open to the council to conclude the proposal was not designated and subject to an EIS. Other cases have focused on the assessment procedure followed by a local council for a particular development. In some cases the court has declared void the council's decision to approve a development because the council has failed to adequately consider, or consider at all, an important aspect of the development e.g. increased traffic caused by a proposed development.[35]

(c) *Judicial Review of Procedural Requirements in Environmental Decision-making (Class 4 Proceedings).* Given the importance of public participation in environmental planning, the Court can review the process by which local councils and the Minister for Planning create zoning instruments (local, regional and state environmental plans). For example, in one case[36] the residents objected to a local environmental plan purporting to enable the local council to develop in a park. The residents succeeded in their action to have the plan declared invalid. The council had failed to advertise the plan sufficiently, as required by the relevant Act. Thus a "technical" deficiency defeated the Council's development intention. While the Council may remake the plan, this time advertising the draft as required, the residents have, through the court action, bought time and gained publicity which may generate sufficient political pressure to have the park left alone. The case illustrates an important characteristic of public interest environmental litigation. Most cases are not "knock outs" but focus political attention on environmental decisions. The political arena, not the legal one, becomes the ultimate arena for public accountability.

(d) *Standing in the Local Council's Shoes-Merit Appeals (Class 1 Proceedings).* In a limited class of cases, known as objector appeals to designated development, third parties can appeal against the local council's decision to approve a class of development. Any person may object and therefore gain an automatic right of a merit appeal to the Land & Environment Court. There have been many court challenges by local conservation groups to local council decisions to approve designated developments such as large mines, quarries and chemical plants. The court has in many of these cases granted development consent but required the development to comply with tougher conditions of development than the local council had required.[37]

An example of one case was a challenge by a local koala protection group, the

Myall Koala and Environment Support Society, against a decision by a local council to approve sand mining in areas of known koala habitat.[38] Koala populations in Australia have been drastically reduced in part because of habitat destruction. The Court did not make the orders sought which were to declare the development consent void but did approve a set of tougher development conditions on the developer.

Innovative Measures to Remove Hurdles to Public Interest Litigants

Because of its specialist nature the Land & Environment Court has established several innovative procedures which place fewer obstacles in the path of genuine public interest litigants.

A legal system which aims to have adequate citizen participation provisions must ensure that access to courts is not severely hampered by procedural rules. For example, costs orders which "follow the event" (the loser-pays rule) so that the loser in court proceedings is liable for their own and the other parties legal costs is very prohibitive of litigation.

Other procedural practices in courts which inhibit public interest litigants include the requirement for undertakings to pay damages when an interim injunction is sought. Security for costs orders can also be prohibitive for community groups and individuals.

Time constraints can also be a major problem for citizen actions. Strict time limits within which court proceedings must be commenced can be difficult to meet for many citizen groups.

Undertakings to Pay Damages. In many courts applicants who wish to obtain an interim injunction are often required to give a personal undertaking to pay damages in the event that they lose the final hearing. The Land & Environment Court has held that no such undertaking is required in genuine public interest cases.[39] The Court would tend to regard these as cases where there appears to be a threat to the environment which the court action is seeking to restrain or remedy.

Security for Costs. Other court orders made against individuals or groups are orders made to provide security for costs. The purpose of these is to have the applicant provide evidence of its means to pay any adverse costs order made against it. The Court has generally not required citizens groups to pay these amounts where they are engaging in public interest litigation.

Reversing the Usual Costs Order. In Australian courts the loser pays under the usual costs rule. This makes litigation for public interest groups seeking to protect the environment very difficult as they are often poorly resourced for this task. In a very few public interest cases the Land & Environment Court has been prepared not to make the usual costs order. One case was a challenge by a member of Actors Equity, the actors union in Australia, against a decision by the state Minister for Planning to lift a permanent conservation order on a large historic theatre under the Heritage Act (1977) NSW. The case was unsuccessful but the court held the matter was in the public interest and the usual costs order would not therefore apply.[40] The order is not made lightly and no similar order was made until 1992 in proceedings challenging a local council decision to approve a multi-storey tower block in a residential area. Although they were unsuccessful the court did not order the residents group to pay costs.[41]

Mediation. Alternative dispute resolution, the phrase used by litigation focused lawyers, includes arbitration, negotiation and mediation. Mediation in particular is being used increasingly to encourage early and satisfactory settlement of environmental disputes. In the Land & Environment Court in NSW certain disputes can be voluntarily referred to mediation before a hearing date is set. The procedure has been received favourably by lawyers and clients alike in numerous cases.

Other states in Australia have adopted differing models for environmental legal disputes. Several have planning tribunals which review local council decisions and have more limited jurisdiction than the Land & Environment Court of NSW.[42]

Other jurisdictions do not have specialist courts to consider environmental law and such issues are dealt with by existing courts. Arguments in favour of specialist courts include that these can develop expertise and jurisprudence in environmental law. This suggests that environmental law is different to other legal areas and requires a different approach to legal enforcement because of its "public" nature. Against this, specialist courts are sometimes viewed as marginal and not part of the judicial mainstream. This possibly leaves them open to subtle political pressure because many court decisions have political consequences where large private and public development is concerned.

6. Resourcing Citizen Action

Despite favourable legal circumstances such as broad standing provisions, volunteer and conservation groups who want to protect the environment are reluctant to engage in court action. Major reasons are inadequate resources as well as the disadvantageous procedural rules in many jurisdictions discussed above.

Legal Aid

There is some help provided in NSW for groups wishing to litigate. The Legal Aid Commission of NSW[43] provides limited funds for appropriate kinds of environmental cases. Legal aid in Australia is a government funded programme aimed at enabling access to the legal system for disadvantaged individuals and groups. The system is often chronically underfunded due to economic constraints on government.

In NSW the Legal Commission established an expert advisory body on environmental cases in 1986 which provides recommendations to Commission officers about which applicants for legal aid should receive a grant under the special public interest environmental law guidelines. It is difficult to satisfy the guidelines. Basically the applicant must show they are acting in the public interest, have reasonable chance of success on the legal grounds and the successful outcome of the case will protect the environment. In 1991 the guidelines were amended to include the consideration of economic interests as relevant to the question of whether a matter is in the public interest. Not surprisingly this created much unhappiness amongst those concerned with environmental protection issues. It also raised squarely the issue of what is "public interest".

No other Australian states provide legal aid on a regular basis or have guidelines for granting legal aid in public interest environmental matters. It therefore occurs infrequently or not at all in most Australian states. There is also provision for legal aid to be granted at the Commonwealth level for public interest cases. Grants under these guidelines are made infrequently. The Mt Etna bat case referred to at page 79 was able to be mounted because legal aid was available from the national government. As legal aid does not cover security for costs orders the applicant was still able to be defeated because of procedural hurdles.

Obtaining a grant of legal aid is very important for most conservation groups who wish to litigate. The grant covers their legal fees and the costs of any experts who are required to give evidence in the proceedings. In the event they lose, a litigant who has legal aid will also have any adverse costs covered by the grant of legal aid. It is of considerable benefit to obtain the grant.

Several cases mentioned earlier were only able to proceed because of the availability of legal aid in NSW. The cases taken by Wendy Jarasius against the Forestry Commission of NSW (pages 85), by Vaughan-Taylor against the local council and a mining company (page 80) and by Nettheim against the Minister for Planning (page 87–88) were all legally aided. The Australian Conservation Foundation was able to take its 1989 action in the Federal Court against the Minister for Resources (page 79) because it was granted legal aid by the Legal Aid Commission of NSW.

Specialist Legal Centres

A specialist independent and non-profit legal centre has also been established in NSW to undertake environmental law work for groups and individuals who are trying to protect the environment. Called the Environmental Defenders Office, it was opened in 1986.[44] Similar bodies exist on a part-time basis only in two other Australian states. Such bodies can only exist in Australia with government assistance, there being few private foundations willing or able to fund such offices. Similar offices exist in Canada, such as the Canadian Environmental Law Association and the Environmental Law Centre in Alberta. Funding for these comes from government and also Law Foundations.

The U.S.A. has several very large public interest environmental law offices such as the Sierra Club Legal Defense Fund, the Natural Resources Defense Council and the Environmental Defense Fund. These groups are about the best funded and largest organisations of their kind in the world. Their existence was made possible to a large extent by the existence of large philanthropic foundations which provide initial and/or on-going funding for these offices. They undertake a variety of work ranging from litigation to lobbying to research and education work.

Reflecting the planning focus of a particular system there are also community planning advisory bodies established which provide assistance to groups and individuals concerned with community planning modelled on similar groups in England.

Free Lawyers

Some lawyers are prepared to act for free in public interest matters. Finding these lawyers is often a random process. In some jurisdictions "pro bono publico" (lawyers acting for no charge) schemes are being considered or have been established.[45]

7. Conclusion

The role of courts and citizen action varies between jurisdictions and will differ between civil law and common law countries. This paper has sought to highlight some of the policy issues which may underlie the role played by the courts as environmental law enforcement institutions and citizen actions in activating those institutions. The significant role that courts can play is shown by focusing on one example of a court where public interest environmental litigation is pursued. A legal system which encourages citizen action through court must also implement other measures to ensure that citizen action is enabled in practical terms. Merely providing a right of action in legislation is insufficient because of the substantial practical impediments to litigation. Further, adversarial litigation is often a blunt tool for protecting the environment. Political, social, economic and administrative tools must not be crowded out with narrow legalism.[46]

The growing debate about environmental issues at the international level will increase substantially the possible scope for citizen action through non-government organisations at the international level. This may impact on national laws and will continue as an important source of environmental policy in the next decade.

Notes

1 In seminal articles such as Christopher Stone's "Do Trees Have Standing" the status of the environment in courts is discussed. As the title to this article suggests the author considers the approach courts could take if the environment or aspects of it could be granted standing in courts without requiring a human plaintiff to act on its behalf.

2 All these nominated categories are areas of civil law. The criminal law is also playing an increasing role in many jurisdictions with the expansion of criminal sanctions against polluters. In some jurisdictions such as Australia and Canada there are now laws which provide for personal criminal liability for company directors whose companies have transgressed pollution laws, e.g., Environmental Offences & Penalties Act 1990 (NSW). The focus of this paper will be the civil aspects of environmental law.

3 See example, Arthurs HW, Without the Law, Administrative Justice and Legal Pluralism in Nineteenth Century England, Toronto, Uni of Toronto Press, 1985, Ch. 1 and 7; Harlow and Rawlings, Law and Administration, 1984.

4 This problem has been recognised by the Australian Law Reform Commission in its report *Access to the Courts: Standing in Public Interest Suits* Discussion Paper No. 4, 1977.

5 *Serrano* v *Priest* 20 Cal 3d25, 11, 568 P. 2d 1303, 1313, 111 Cal Rptr 315, 325 (1977).

6 Examples include *Into the Future: Environmental Law and Policy for the 1990s* Environmental Law Centre (Alberta) Society 1990 which includes a chapter "Public Interest Litigation in Canada" A. Roman and M. Pikkov. *Environmental Law in Australia* G. Bates, Butterworths, 1992. There are numerous journal articles in this area also. Bryden *Public Interest Intervention in the Courts* (1987) 66 Canada Bar Review 490; Capelletti *Vindicating the Public Interest through the Courts: A Comparativist's Contribution* (1976) 25 Buffalo Law review 643 (U.S.A.); O'Connor *Reflections on the Preclusion of Judicial Review in England and the U.S.* (1986) 27 Wm & Mary Law review 643 (Eng); and from the Netherlands Groenendijk *Litigation, Politics and Publicity: Public Interest Law or How to Share the Burden of Change* (1985) 14 Anglo-American Law review 337.

7 Robinson D *Pollution Control and Public Interest Litigation*, Environmental Defenders Office, Sydney, Australia, 1991 pp. 2–5.

8 Seed J et al., "Thinking Like a Mountain: Towards a Council of All Beings" London, heretic Books, 1988 at p. 23.

9 One of the best known examples is the U.S. Environment Protection Authority.

10 There are now many examples from several jurisdictions where provisions enable citizen participation, e.g., section 123, Environmental Planning & Assessment Act 1979 (NSW).

11 Freedom of Information legislation exists in many jurisdictions. The strength of the legislation differs between jurisdictions however. In Australia two states out of six have such legislation. It also exists at the Commonwealth level. Each statute differs in the scope of the government legislation it makes available to the public.

12 Greve, M.S., *The Non-Reformation of Administrative Law: Standing To Sue and Public Interest Litigation in West German Environmental law* (1989) 22 Cornell Int. Law Jl 197 at p. 198.

13 The Rose Theatre case, *R* v *Sec of State for the Environment ex parte Rose Theatre Trust Company*

in the U.K. demonstrates this issue. The court held that the applicants did not have standing to sue for judicial review of a decision not to place a heritage order over a building by a Minister on the basis that if no individuals have standing then no group formed by concerned individuals could have standing. The statute required that "sufficient interest" be shown.

[14] Cripps, C.J., "The Land & Environment Court" College of Law, Sydney, NSW, June, 1990.

[15] The law of standing will vary depending on the composition of the court deciding the issue. A more conservative U.S. Supreme Court has recently decided in *Lujan* v *Defenders of Wildlife* 110 S.Ct 3177 to deny the applicant standing. The decision is considered to limit and contract the previously expansive standing provisions applied under the U.S. administrative Procedures Act whereby it was relatively easy to show the applicant had interests which were affected and therefore had standing.

[16] *Australian Conservation Foundation* v *Commonwealth* (1980) 28 ALR 257.

[17] *Australia Conservation Foundation* v *Minister for Natural Resources* Unreported, Federal Court of Australia, No. G326, 1989.

[18] Export (Unprocessed Wood) Regulations under the Customs Act (Cth).

[19] Administrative Decisions (Judicial Review) Act (Cth) 1977.

[20] Administrative Decisions (Judicial Review) Act (Cth) 1977 s. 3(4) provides that a person aggrieved includes "a person whose interests are adversely affected by the decision" s3(4)(i).

[21] *Queensland Speleological Society* v *Queensland Cement Pty Ltd.*

[22] It was a pity that the High Court did not use the opportunity to relax the strict standing precedent laid down in the 1980 judgment in *ACF commonwealth.*

[23] *Vaughan-Taylor* v *Kempsey Shire Council & David Mitchell-Melcan* (1991) 73 LGRA 366.

[24] Other legislation which provides broad standing in NSW includes the National Parks & Wildlife Act, the Heritage Act and the Environmentally Hazardous Chemicals Act. New pollution law, the Environmental Offences & Penalties Act was introduced in 1989 which opens up the possibility for standing for citizens to a greater degree than was previously possible for concerned citizens.

[25] The NSW Law Reform Commission is currently investigating wider use of civil enforcement of environmental laws.

[26] Environmental Offences & Penalties Act (1989) NSW, Australia.

[27] Section 13, Environmental Offences & Penalties, provides that any person can take prosecution action (without requiring permission from the state government) but must satisfy the court as a preliminary issue that they can prove their case to a *prima facie* standard and are acting in the public interest.

[28] Section 25, Environmental Offences & Penalties Act requires that the court must be satisfied that the proceedings are not an abuse of process; that there is a real or significant likelihood that the requirements for the making of an order under this section will be satisfied; and the proceedings are in the public interest before giving leave for the proceedings.

[29] *A.J Brown* v *Environment Protection Authority*, LEC No 40093, commenced June 1992.

[30] There is reference in other fields of law to "class actions" where a class of people who have been affected by, say, a defective product take action for damages as a group. This type of case is prevalent in the U.S.A. In Australia class actions in the environmental law field are not taken probably because damages for injury are not sought but rather declarations are sought that the law is not being complied with. Actions involving several people are described as "representative" when one person can take action on behalf of a group of people.

[31] Land & Environment Court Act 1979 (NSW).

[32] *Jarasius* v *Forestry Commission of NSW* (1989) 69 LGRA 156.

[33] *Hannan Pty Ltd* v *Electricity Commission of NSW (No 1)* [1983] 3 NSWLR 282.

[34] *Malcolm on behalf of Marylands Residents Group* v *Newcastle City Council* (1991) 73 LGRA 356.

[35] *Parramatta Shire Council* v *Hale* (1982) 47 LGRA 319.

[36] *Canterbury District Residents & Ratepayers Association* v *Canterbury Municipal Council & Minister for Planning.*

[37] *Bignall* v *Greater Taree Shire Council* LEC No 10293/86 concerning a large quarry development opposed by a local residents group.

[38] *Myall Koala and Environment Support Society* v *Great Lakes Shire Council* LEC No 10096/90.

[39] There are several cases in NSW including *Ross* v *State Rail Authority, Nettheim* v *Minister for Planning* LEC No 40086/88.

[40] *Nettheim* v *Minister Administering the Heritage Act & Anor* LEC No 40086/88.

[41] *Darlinghurst Residents Association* v *Elarossa Investments Pty Ltd* (no 3) LEC No 40225/91.

[42] In Victoria there is a Planning Appeals tribunal which considers merit appeals on planning issues. Queensland has the Local Government Court, which has more limited jurisdiction than the Land & Environment Court of NSW.

[43] A statutory body established by legislation, the Legal Aid Commission Act (1979) NSW, the Commission is supposed to be independent of government in deciding legal aid policy.

[44] The Environmental Defenders Office is a non-profit company incorporated in NSW. It has a Board of Directors and employs several lawyers as well as other staff.

[45] The NSW Law Society and the Public Interest Advocacy Centre in Sydney, NSW, are establishing a clearing house for pro bono work by lawyers so that individuals or groups with public interest concerns including environmental issues can find lawyers able to act for free in litigation.

[46] See Ayres I and Braithwaite J *Responsive Regulation: Transcending the Deregulation Debate* New York 1992; Braithwaite J *The Future of Regulatory Enforcement in Australia* Australia Institute of Criminology, Canberra, Conference Proceedings 1992.

6. Addressing Uncertainty
Law, policy and the development of the precautionary principle

JAMES CAMERON[1] and WILL WADE-GERY[2]
[1]*SOAS, University of London; Life Member, Clare Hall, Cambridge University; Visiting Professor, College of Europe, Bruges, Belgium*
[2]*J.D. Candidate, Yale University, Yale Station, Connecticut, U.S.A.*

1. Introduction

Human society is preoccupied with change. Threatened and stimulated, we see the inevitability of change as a challenge upon which our survival as a species depends as well as our apparent supremacy. In this stressful and complex world certainty recedes with knowledge – the more we know the less we know and the harder it is to be told. In world of politics and law, rational decision-making is customarily based on varying, degrees of particular forms of certainty associated with the notion of proof. When we are uncertain we politically paralysed. Too often, our experience in matters relating to the environment indicates that when we are certain we are impotent – it is too late to repair the damage.

The precautionary principle carries a structure of ideas to itself which enable us to take decisions which seek ecological balance, for the benefit of our human society. The principle offers guidance, within the embrace of the law, as to how we might interfere least, or least damagingly, in the ecosystems which support life on earth.

The precautionary principle provides the philosophical authority to take decisions in the face of uncertainty. To take what we know about the balance of nature as a reason for regulatory action and to remove the conceptual barrier to action caused by our ignorance of nature.

The precautionary principle has a number of strengths that have assisted it's quite spectacular advance into the consciousness of the world's environmental decision makers. It is general in character – offering interpretive nuances to many cultures, languages and political/legal systems. It is symbolic of the need (and widespread political will) for change in human behaviour towards the environment that sustains our existence. It both challenges science and respects fundamental principles of ecology. It is both heretical and traditional. It has garnered grass roots support – take as an example a vernacular representation created by Greenpeace:

Bruno Dente (ed), Environmental Policy in Search of New Instruments, 95–142.

"If in doubt don't pump it out!"

This paper argues that the precautionary principle of environmental regulation now constitutes a general principle of international environmental law. As such, it has informed and should increasingly continue to inform the effective formation and implementation of a range of policies instrumental in the protection of the environment. The precautionary principle, it will be argued, represents the correct legal and policy response to the scientific uncertainties surrounding current and potential environmental degradation. The principle can serve simultaneously both as a general international cultural standard for environmental protection and as a determinant or guide for more specific policy instruments designed to effect that protection.[2]

The paper proceeds in four main stages. First, the development of the principle at the level of international cooperative discourse is reviewed. The conceptual core of the principle is defined and subsequently elaborated. This definition, it will be suggested, can serve to resolve some of the unnecessary ambiguities and confusions concerning the principle, both at a philosophical level and in terms of its more practical implications. Second, and following from this review, the development of the principle at the level of international law is examined. Third, we argue both for a recognition of the significance of the principle and for its further development into a procedural realm. These latter arguments are based on what we feel to be the principle's effectiveness in overcoming traditional problems of international cooperation (or rather non-cooperation) and in supplying a range of policy instruments to address widely-acknowledged environmental concerns. This range will be assessed and some possible additions to it considered.

2. The Discourse of Precaution: A Precautionary Assessment

2.1. *Explicit Precautionary Language*

The earliest explicit references to the precautionary principle were contained in domestic West German legislation in the early 1970s.[3] At the international level, the rise of explicit precautionary discourse has only been evident in the latter part of the 80s. But that rise has been conspicuously meteoric.[4] By 1990, the U.N. Secretary-General was able to report that the precautionary principle "has been endorsed by virtually all recent international forums."[5]

The earliest international agreement to employ explicitly precautionary discourse is widely accepted to be the Ministerial Declaration of the Second International conference on the Protection of the North Sea, issued in London in November, 1987.[6] In this, the Ministers accepted, *inter alia*, that:

> . . . in order to protect the North Sea from possibly damaging effects of the most dangerous substances, a precautionary approach is necessary which may require action to control inputs of such substances even before a causal link has been established by absolutely clear scientific evidence.[7]

Signatories to the Baltic Sea Declaration stated their "firm intention" to:

> Apply the precautionary principle, i.e., to take effective action to avoid potentially damaging impacts of substances that are persistent, toxic and liable to bioaccumulate even where there is a lack of full scientific certainty to prove a causal link between emissions and effects.[8]

An exhaustive listing of references to the precautionary principle is beyond the scope of this paper. However, several of the more recent examples are laid out below in order to give a sense of the subtle variations in the language employed, and of the range of fora to have endorsed the principle in explicit terms.

2.1.1. *European Community Environmental Legislation*

The proposed revision to the Treaty of Rome as agreed at Maastricht now states that:

> The Community policy on the environment . . . shall be based on the precautionary principle and on the principles that preventive action should be taken, that environmental damage should as a priority be rectified at source and that the polluter should pay. Environmental protection requirements must be integrated into the definition and implementation of other Community policies.[9]

Also worth noting are two directives on Genetically Modified Organisms (GMOs) approved by the European Council in 1990.[10] These stipulate with varying degrees of specificity a range of precautionary procedures that must be followed in the contained use and deliberate dissemination of GMOs into the environment. These procedures are designed to protect both the environment and its human inhabitants from the risks and uncertainties associated with GMOs.

2.1.2. *Sustainable Development*

The 1992 Rio Declaration on Environment and Development states:

> Where there are threats of serious or irreversible damage, lack of full scientific certainty shall damage, lack of full scientific certainty shall not be used as a reason for postponing cost-effective measures to prevent environmental degradation.[11]

2.1.3. *Global Warming*

The 1992 U.N. Framework Convention on Climate Change states that:

> The Parties should take precautionary measures to anticipate, prevent or minimise the causes of climate change and mitigate its adverse effects. Where there are threats of serious or irreversible damage, lack of full scientific certainty should not be used as a reason for postponing such measures, taking into account that policies and measures to deal with climate change should be cost effective so as to ensure global benefits at the lowest possible cost.[12]

2.1.4. *Radioactive Waste*

Agenda 21, agreed at the 1992 Rio conference, recommends in relation to radioactive waste that States should not:

> promote or allow the storage or disposal of high-level, intermediate level and low-level radioactive wastes near the marine environment unless they determine that scientific evidence, consistent with the internationally agreed principles and guidelines, shows that such storage or disposal poses no unacceptable risk to people and the marine environment or does not interfere with other legitimate uses of the sea, making, in the process of consideration, appropriate use of the concept of the precautionary approach;[13]

2.1.5. *Ozone Protection*

Although the original Vienna Convention of 1985 and its subsequent Protocol on Substances that Deplete the Ozone Layer, which was established in Montreal in September, 1987, do not use explicitly precautionary language, the protocol has since been amended to include this new terminology. Thus the June 1990 London Amendments to the Montreal Protocol state that:

> [The Parties to this Protocol are] determined to protect the ozone layer by taking precautionary measures to control equitably total global emissions of substances that deplete it, with the ultimate objective of their elimination on the basis of developments in scientific knowledge, taking into account technical and economic considerations and bearing in mind the developmental needs of developing countries.[14]

This amendment serves as but one example of a growing – and sometimes controversial – phenomenon, namely the inclusion of precautionary terms into previously established environmental regulatory regimes.[15] The terms of this controversy are themselves extremely revealing. Objectors to the inclusion of specifically precautionary discourse have in some cases based their objections on the apparent redundancy of such language; precautionary standards are, it is claimed, already embodied in the established standards of environmental protection, and are therefore unnecessary to the correct functioning of the regime in question.[16] This line of objection, we will argue, is both true and false. It is true inasmuch as the concept of a precautionary approach to environmental regulation predates the development of explicitly precautionary language. However, it is false inasmuch as it claims that this predating negates the significance of the precautionary principle as an internationally recognized basis to the development of environmental law and policy. In point of fact, the very vociferousness of the objections point to this fallacy; truly redundant developments in legal or policy discourse are not strenuously blocked in this fashion. In some cases, where such blocking is practised by a state or group at their demonstrable political expense, we have prima facie evidence of the political importance of the idea being blocked. These arguments are further developed below in Section 3.1.

2.2. *Conceptual Elements of the Principle: Moving Beyond Explicit Discourse*

Although they cover a wide range of international environmental concerns and decision-making (or decision-influencing) bodies, the explicit references noted above only partially indicate the extent of the principle's adoption. Precautionary approaches to environmental law and policy are also implicit. To analysis these implicit references to the principle obviously requires some elaboration of the conceptual core of the precautionary principle.

Such an elaboration has been assumed by some commentators to be difficult to effect. Gundling, for instance, has described the "principle of precautionary action" as an "elusive concept" particularly "in an international context where problems of concept are increased by linguistic ambiguities."[17] Moreover, this characteristic of elusiveness is taken to imply that the principle cannot be said to qualify as a general principle of customary international law.[18] Thus Handl, noting the various definitional ambiguities of the concept, states unequivocally that "at present, the precautionary principle is not a term of art."[19]

Moreover, even the cursory review of recent explicit references given in 1.1 above, demonstrates considerable linguistic variation.[20] But these truths do not validate the claims of Gundling and Handl. The precautionary principle does have a conceptual core. It is not, to use Gallie's well-known formulation, an "essentially-contested concept." Contestations, in practice, centre around the degree and effect of the principle's adoption in law and policy. The many references in international legal materials do not directly question the essence of the principle.

2.3. *Defining the Core of the Principle*

Much of the confusion surrounding the principle's interpretation relates to its distinction from more traditional preventive standards of environmental prevention. Put simply, the precautionary principle both in its conceptual core and its practical implications, is preventive, but not all preventive standards are precautionary. More exactly, any particular preventive standard may be either non-precautionary or precautionary in various degrees. By contrast, any given precautionary standard may be preventive in various degrees, but it cannot be unpreventive.

Consider the International Maritime Organization International Convention on Oil Pollution Preparedness, Response and Cooperation, established in 1990.[21] According to the terms of the Convention, Parties to it, "recognizing the serious threat posed to the marine environment by oil pollution incidents involving ships, offshore units, sea ports and oil handling facilities," noted that they were "mindful of the importance of precautionary measures and prevention in avoiding oil pollution in the first instance."[22] Despite the presence of precautionary language here, there is nothing precautionary about the standards set. The threats posed to the marine environment are clear. Measures are being taken to prevent such known threats from being realized. The certainty of the environmental damage that would result from a failure to adhere to those standards, means that the Convention is not precautionary. It is however, unquestionably preventive in its intention. Oil pollution is a known environmental hazard; measures are to be put in place to prevent that hazard being realized. Parties are aiming to reduce the risks of known, that is certain, environmental hazards.

The terms of the convention may be contrasted with those of the Conference for the Protection of Coasts and Waters of the NorthEast Atlantic Against Pollution Due to HydroCarbons or Other Harmful Substances. The Conferences Final Act declared, inter alia, the need for measures designed to prevent discharges of "other harmful substances," where the latter were defined as substances "the release of which into the marine environment may lead to injury to human health, to ecosystems or living resources, or to the coasts or related interests of the Parties."[23] Here the risks being reduced are of an unknown nature. It is unclear what environmental damage the release of these "other harmful substances" into the marine environment would cause. The standard set is obviously preventive in intent, since it clearly seeks to prevent environmental damage, but it is also precautionary, in that the standards set are a response to the uncertainty surrounding the environmental effects of particular discharges. The crucial term, of course, is 'may.'

This example provides us with the key element of the conceptual core of the precautionary principle. A lack of certainty about cause and effect relationships, or the extent of possible environmental harm, does not legitimate delaying the imposition of some kind of regulatory mechanism over the activity in question.

The precautionary principle stipulates that where the environmental risks being run by regulatory inaction are in some way (a) uncertain but (b) non-negligible, regulatory inaction is unjustified.[24]

This conceptual core allows us to answer in the affirmative Nollkaemper's well-phrased inquiry: "The question is whether or not the principle distinguishes itself in the already dense normative scenery in this field?"[25]

2.4. *Beyond the Conceptual Core*

This conceptual core does leave open the answers to three questions. This openness, it may be reasonably suspected, is what accounts for what is incorrectly seen as the definitional ambiguities of the precautionary principle. These three questions, it must be stressed, do not address the essence of the principle. What they do is augment the distinction in kind already drawn, with a series of differences in precautionary degree.

The three questions prompted by the core principle may be given as:

(1) What counts as a non-negligible environmental risk?
(2) What regulatory action is justified?
(3) How should the answers to questions (1) and (2) be determined?

2.4.1. *Non-negligible Environmental Risk*

Question (1) above has occasioned some confusing answers. Gundling, for instance, has sought to define the precautionary principle as "more than the prevention of risk." He elaborates:

> Precautionary action requires reduction and prevention of environmental impacts irrespective of the existence of risks . . . [It] also requires action even if risks are not yet certain but only probable, or, even less, not excluded.[26]

In this case, non-negligible environmental risk arises in all cases of environmental impacts. This is clearly too broad an understanding of the precautionary principle. Human activity as a matter of course entails environmental impacts. Not all such impacts should therefore be mitigated by the imposition of environmental regulation. As Cameron and Abouchar have noted: "When evaluating our responsibility in regard to the precautionary principle, we must recognize that human pollution is inevitable."[27] In essence, not all environmental risks are non-negligible and a failure to recognize this fact introduces a utopian element into the precautionary principle, which cannot be sustained.

The precautionary principle thus requires criteria to determine a threshold of non-negligible environmental risk. Below that threshold, no regulation is required. Above the threshold, regulation is required. The question of how such a threshold should be determined is part of question (3) noted above. The issue of determination is therefore examined in Section 1.4.3 below.

There is however a crucial difficulty surrounding the precautionary principle that, since it relates to the notion of risk, should be addressed at this stage. This difficulty is evident, for instance in a 1991 OECD study on the precautionary principle, "Dealing with Uncertainty," which notes that:

> some would like to see [the principle] applied to any substance liable to cause pollution, while others consider that it should be applied only to substances that have adverse effects, or to substances whose emissions actually do create a pollution risk.[28]

But if a substance is liable to cause pollution or is known to have adverse effects, then controlling or regulating such substances is a purely preventive measure. There is no element of uncertainty involved, and the precautionary principle cannot apply. Thus the first two types of substances listed by the OECD above (those that are "liable to cause pollution" and those that "have adverse effects") cannot be regulated by reference to the precautionary principle. The third type of substances noted (those that "actually do create a pollution risk") require further elaboration.

The risk of confusion here arises from the apparently unsteady distinction between risk and uncertainty. "Risk" in this and other contexts is usually defined as the amalgam of the probability of an event occurring and the seriousness of the consequences should it occur. Thus a high-risk strategy is one that either combines a relatively high probability with relatively innocuous consequences or one that combines a relatively low probability with relatively serious consequences. For instance, the 1990 UN ECE Draft Code of Conduct on Accidental Pollution of Transboundary Inland Waters (which relates to the protection of transboundary bodies of water, including groundwater, against accidental pollution resulting from hazardous activities), defines a "hazardous activity" as "any activity which by its nature involves a significant risk of accidental pollution of transboundary inland waters,"[29] and then further defines risk as "the combined effect of the probability of occurrence of an undesirable event and its magnitude."[30]

Regulatory standards that prevented hazardous activities as noted above would indeed be precautionary, but they would not be uniformly so. The precautionary extent of the standards would depend on the relative certainties embodied in the risk or risks in question. If both the probability of accidental pollution and the magnitude of the consequences of that pollution are known, the standards would be relatively unprecautionary, precisely because the level of uncertainty involved is relatively low. High risks do not necessarily entail high levels of uncertainty. However, if the probability and magnitude are relatively unknown, because, for instance, it is not known what cause and effect relationships are involved, or exactly what the nature of the involved causal relationships is, then the standards would be precautionary because of the relative uncertainties involved.

This points to an interesting phenomenon. Some eco-systems and human communities may, because of their physical circumstances, face greater risks than others from the same (global) phenomenon. In this way, if an international regulatory standard were to be introduced, it might be less precautionary for some communities than for others. The example of the UN Framework Convention on Climate Change may be informative in this regard. As is widely known, the scientific evidence relating to the cause and effect relationships between the emission of greenhouse gases, accelerated climate change, and damage to the environment, is in some respects notoriously inconclusive.[31] However, the probability of change in the climate system given current level of emissions is known to be high. At the same time, probabilities of more substantial climate changes can only be speculatively arrived at: the various feedbacks are not completely understood. A high level of uncertainty pervades such calculations. Some parts of the globe, particularly low-lying small islands, would be very seriously and possibly fatally affected by the consequences of small changes in temperature, especially in respect of coral bleaching and sea-level rise.[32] Efforts to prevent a rise in sea-level are therefore for these island communities barely precautionary; the threat to these islands is known with considerable certainty. Addressing the Alliance of Small Island States in February 1992, the Minister for Planning and External Relations of the Republic of Seychelles, observed:

> We are the front line states. If the IPCC Report is correct . . . then AOSIS members are doomed to ecological, social and economic catastrophe within the foreseeable future. Some will even disappear.[33]

But the predicted, if disputed, change in temperature has less clear implications for other communities. If the feedbacks of climate change are self-reinforcing, then the effect could also (eventually) be similarly serious.[34] If not, then the effects might be negligible. Scientific evidence is not enough to establish relative certainty in this regard. For these other communities, any standards set to prevent these possibilities would be relatively precautionary.

This point cannot be overstated. Environmental interdependence dictates that the implementation of a vital preventive standard in one part of the planet is contingent upon the adoption of precautionary standards elsewhere. In other words, the physical survival of certain communities is already dependent on other communities behaving in a precautionary fashion.

2.4.2. *Justified Regulatory Action*

Assuming a determination that the threshold of non-negligible environmental risk has been reached or surpassed, what regulatory standards are justifiable? The 1989 Nordic Council's Conference on the Pollution of the Seas points to the range of variables that are relevant in answering this question:

> . . . the need for an effective precautionary approach, with that important principle intended to safeguard the marine ecosystem by, among other things, eliminating and preventing pollution emissions where there is reason to believe that damage or harmful effects are likely to be caused, even where there is inadequate or inconclusive scientific evidence to prove a causal link between emissions and effects.[35]

In this case, the precautionary principle is invoked for the advancement of a particular end (the safeguarding of the marine ecosystem), and this end is to be attained by particular means (the elimination and prevention of pollution emissions). Thus, question (2) noted above prompts two further questions: first, what end is the precautionary principle designed to realize, and second (and to some extent following from this), what means are justifiably employed to realize (or approach the realization of) these ends?

2.4.2a. *Ends of the Principles*. Not all answers to the first question are as clear-cut as that provided by the Nordic Council. There, the principle is to be employed to secure a particular environmental objective, the safeguarding of an ecosystem.

Similarly, the United Nations General Assembly Resolution on the World Charter for Nature, which contains several precautionary stipulations, states in its preamble its conviction that:

> Every form of life is unique, warranting respect regardless of its worth to man . . .[36]

This suggests that its endorsement of the precautionary principle (which is implicit only) is designed partly to protect biodiversity as an end in its own right. Negative as well as positive environmental objectives are sometimes given. Thus, a 1989 Resolution of UNEP's Governing Council urged precautionary measures on the grounds that:

> waiting for scientific proof regarding the impacts of pollutants discharged into the marine environment may result in irreversible damage to the marine environment and in human suffering.[37]

Avoidance of irreversible damage to the environment is a commonly stated end of the precautionary principle. In some more extreme cases, the negative end of avoiding irreversible damage becomes the positive end of ensuring physical survival. Thus the Alliance of Small Island States (AOSIS) has stated bluntly:

> For us the precautionary principle is much more than a semantic or theoretical exercise. It is an ecological and moral imperative . . . . We do not have the luxury of waiting for conclusive proof, as some have suggested in the past. The proof, we fear, will kill us.[38]

Other invocations of the principle have declared their end to be the securing of sustainable development (the Bergen Declaration is the most widely known example). Still others either augment or substitute the terminology of sustainable development by referring to some combination of environmental ends (both for present and future generations) and economic objectives.[39] Frequently these economic objectives are only implicit; their explicit annunciation is confined to qualifying environmental ends with statements to the effect that means to these ends must take economic factors into account.[40]

This range of ends provides a further basis for scaling the precautionary aspect of any given international environmental regime. The more emphasis placed on environmental ends in their own right, and the less placed on qualifying economic ends, the more precautionary the regime in question. Another way to say this is that in cost-benefit terms, the precautionary principle attributes a very high cost to regulatory inactivity in the face of uncertainty regarding the possible environmental impact of activities deemed to carry a non-negligible environmental risk. Conversely, regulatory activity in these same circumstances carries an inherent benefit, independent of and prior to any scientifically conclusive demonstration of the environmental impacts involved.

The no-regrets doctrine of environmental regulation expounded by the Bush Administration and the G-7 Houston Summit demonstrates an entirely different relationship between uncertainty and these same costs and benefits. Here, the correct response to these same uncertainties is required to be precautionary in an entirely different fashion. Because uncertainties mean that the costs and benefits of regulation cannot be scientifically known now, the costs of regulatory inactivity should be assessed as low, while the benefits of regulatory activity are deemed similarly insignificant. The only basis for regulatory standards is certainty. Thus measures taken ostensibly "in response to uncertainty" must be "justified in their own right", i.e., by their demonstrated capacity to realize other objectives. Thus the Houston summit declaration noted that "in the face of threats of irreversible environmental damage, lack of full scientific certainty is no excuse to postpone actions which are justified in their own right."[41]

Thus a no regrets policy towards greenhouse gas emissions would seek to impose controls on such emissions only in as much as those controls realize in a scientifically-certain manner other ends, such as energy efficiency. The terminology is obviously confusing in this area, since both approaches are cautions and seek to avoid regret. The difference in conceptual and practical terms is, however, very clear. The no regrets doctrine seeks to avoid regretting regulatory action because of the uncertain consequences of taking such action. The precautionary principle seeks to avoid regretting regulatory inaction because of the uncertain consequences of such inaction.

2.4.2b. *Means*. These same points apply to the means employed to implement the declared ends of a precautionary approach. These means are in essence intermediary ends rather than specific procedures of compliance with the requirements of precautionary action. The Nordic Council's report as noted above aims for the "elimination and prevention" of emissions. Other regimes list slightly different subsidiary ends, such as the use of best available technology to control potentially damaging environmental impacts.[42] Still others qualify minimization requirements with economic language.[43] As we have seen, this kind of language does not imply a non-recognition of the precautionary principle. Precaution, both in essence and degree, will depend on the relative costs assigned to regulatory (in)action in the face of uncertainty.

The precautionary principle thus endorses, or arguably, requires "worst-case" environmental planning. Costs of regulatory action are assumed to be low, and the benefits of regulatory action high, because "the worst" is assumed about possible environmental outcomes. This aspect of the principle will be returned to in Section 3. For now, it is sufficient to note that significant environmental standards are more likely to be established at the international and domestic levels, if worst-case planning becomes normatively established. This is for a simple reason: the worse possible outcomes are assumed to be as a matter of course, the larger and more determined will be the constituency demanding these kinds of standards.[44]

Nevertheless, as O'Riordan has noted: "There are some dangers in getting too carried away into the application of precaution at any cost."[45] In particular, high cost regulatory actions, if subsequently shown to be unnecessary, will tend to generate economic waste and political embarrassment, thus reducing the force of subsequent precautionary arguments. The precautionary principle can, however, incorporate a response to such dangers without ceasing to be precautionary. This incorporation can take two forms.

Taking O'Riordans remarks as a structure for analysis it can be argued that the principle requires for its application a cost-assessment procedure. For its application to be maximized in practical terms, these possible costs of over responding to uncertainty should be taken into account by that procedure. Worst case planning does not have to proceed in an entirely compartmentalized fashion. If such planning is conducted at a holistic level, then by the terms of the principle the potential costs of over-reacting must figure in the precautionary calculus of costs and benefits.

Additionally, the precautionary principle mandates regulatory action in response to uncertainty about environmental impacts. There is thus an element of the principle that overtly admits the possibility that any particular regulatory measure may overstate the dangers involved. But the principle also claims that it is better to run the risk of overstatement than understatement. Since these elements of the principle are overt, the potential political costs of overstatement may themselves be overstated. An insurance analogy can clarify this point. The precautionary principle can be thought of as an environmental insurance policy taken out by the environmentally risk adverse.[46] Those paying insurance premiums do not tend to feel that they didn't get their money's worth merely because the insured-against event did not come to pass. The value obtained by an insured is in most cases the insurance itself. Similarly, precautionary environmental standards have value not just because of actual damage averted, but because of their direct insurance value. This latter is not affected by the subsequent course of events.

2.4.3. *Determination of Thresholds and Responses to Uncertainty*

The above sections allow for the question of determination to be rephrased in more specific terms. We can now ask (a) how the threshold of non-negligible environmental risk should be set, and (b) how the relative cost of regulatory inaction in the face of uncertain environmental impacts should be assessed. General precautionary answers to these questions have already been provided. They are, as above (a) low, and (b) high. The question therefore becomes specifically (a) how low, and (b) how high?

The precautionary principle requires answers to these questions to be determined politically by interested parties. Since, by definition the scientific evidence is inconclusive, determinant answers to (a) and (b) cannot be made scientifically. Particular environmental policies must, as O'Riordan has argued, "be accounted for in forms that are more overtly judgemental."[47]

The precautionary principle thus makes a virtue out of necessity. It accepts a non-scientific basis for decision-making, and urges that it incorporate as many interested parties as possible. At the international level, this means that the determination of environmental law and policy must involve more than state actors.[48] Thus O'Riordan notes:

Precaution . . . tests political institutions in that they are forced to regulate public and private affairs without recourse to the authority of formal science. This permits aggressive entry into political decision making both by national and international pressure groups seeking to exploit scientific uncertainty and political indecisiveness.[49]

The precautionary principle thus allows for the democratization of international environmental regimes. By explicitly noting the limits of scientific determination, it legitimates a public political determination of these questions.

The question remains, however, of how much this political process of evaluation can supersede scientific evidence. International precautionary statements are not uniform in this regard. The Final Ministerial Declaration of the Third International Conference on the Protection of the North Sea stated that:

[The ministers pledged to] continue to apply the precautionary principle, that is to take action to avoid potentially damaging impacts of substances that are persistent, toxic and liable to bioaccumulate even where there is no scientific evidence to prove a causal link between emissions and effects.[50]

In this case no scientific evidence of a causal connection is required for precautionary members (and thus political judgements) to come into effect. In other cases the precautionary domain is more narrowly drawn. The Second ICNS Ministerial Declaration states that a "precautionary approach . . . may require action to control inputs" of potentially damaging substances "even before a causal link has been established by absolutely clear scientific evidence."[51] The implication is that inputs cannot be controlled unless some scientific evidence of a less than absolutely clear nature is present. This regime is therefore less precautionary than the Third ICNS Ministerial Declaration.

The question of the precise relationship between science and the precautionary principle has been a matter of some controversy. In a widely-noted piece, Gray has argued that "acceptance of the precautionary principle is entirely an administrative and legislative matter and has nothing to do with science."[52] This seems unlikely for a variety of reasons.

First, one of the major factors behind the rise of the precautionary principle both nationally and internationally has been the public perception of scientific inadequacy in addressing questions of environmental degradation. Public and heated scientific disagreements and failings have added to the political force of the precautionary principle. The principle has achieved public articulation as a consequence of the development of a Kuhnian "pre-revolutionary" situation, in which one "tradition of scientific practice . . . is felt to have gone badly astray."[53] As Jackson and Taylor note, the scientific community entirely failed to predict certain phenomena such as ozone-depletion, forest die-back; phenomena that were predicted in general terms, such as global warming, soil and lake acidification, and the eutrophication of coastal waters, were not sufficiently understood to establish points of onset or rates of change; and known environmental degradations have in many cases not been complemented with knowledge of their causes and effects.[54]

Second, and as a matter of theory, it is unclear that science provides an adequate basis for environmental decision making in that it cannot eliminate uncertainties. The very nature of the scientific endeavour therefore provides an argument for the application of the precautionary principle. The OECD, for instance, has noted that scientific theory is itself "fraught with uncertainty."[55] Similarly, the Bergen "Joint Agenda for Action" of May, 1990 in discussing sustainable industrial activity called for environmental regulation "based on scientific foundations recognizing the limits of science." The Agenda also ascribed to the scientific community a"compelling obligation . . . to explain clearly the nature of scientific knowledge and the certainty of uncertainty."[56]

In what sense can scientific practice be regarded as inherently uncertain in this particular field? The main problem in this regard is the difficulty of establishing cause-and-effect relationships in highly complex eco-systems interactions.[57] In the case of North Sea pollution for instance, the marine eco-system is too complex to establish clearly-delineated critical thresholds for particular environmental inputs. Thus Sperling:

> Tests can only reflect a very small, and not necessarily representative part of the occurrences in the marine ecosystem. Observations in the field, on the other hand, provide clear results only in the event of very drastic phenomena. For these reasons, forecasts of risks are extremely uncertain.[58]

But the limitations of statistically-obtained experimental evidence of causality, substantial though those have been in many cases (as later evidence has shown), are minor compared to the limitations of theoretically-advanced evidence. Establishing causal relationships in the area of climate change, for instance, necessarily depends on theoretical modelling; hypotheses will not be testable empirically until putative effects may have become irreversible. Moreover, this theoretical modelling or simulation must cover the range of interactions between the planetary eco-system and the human socio-economic system. Forecasting an interaction so substantial and complex is inevitably an uncertain business. As Skolnikoff notes:

> Complex systems that are incompletely understood may have unsuspected non-linear responses when subjected to unusual stress . . . Surprises such as the Antarctic ozone hole cannot be rules out on the basis of current knowledge.[59]

Third, all scientific assessment of environmental damage, either actual or potential, is dependent on non-scientific, subjective assessments of what constitutes environmental damage, or degrees of acceptable or unacceptable environmental damage. As Greenpeace have claimed, all claims for scientific assessment as a uniquely objective form of policy guidance "fail to recognize that all approaches to control of wastes in the environment must involve judgements on the value and acceptability of changes in the environment, which have a cultural, political or economic basis."[60] This is not of course an argument for the precautionary principle itself, but it does effectively counter one type of argument that, as Greenpeace notes, is not infrequently deployed against calls for precautionary policies.

It should be stressed that the controversy that surrounds the relationship of science and technical expertise to wider more democratic publics is not one confined to the precautionary principle; the debate is long-standing and far-reaching. The tension between democracy and reason is evident, for instance, in Dewey's "The Public and its Problems,"[61] which "espoused an idealistic vision of neighbourly discourse as the foundation of all political decision, but also argued that most of the concerns of the modern state are technical matters, best left to the experts to resolve on the basis of the facts."[62]

Finally, it is worth noting that there have been some efforts to obtain explicit international recognition of the inherent nature of scientific uncertainty pertaining to the effects of environmental inputs. For instance, at the Bergen Conference, the Swiss delegation sought to include in the Final Ministerial Declaration a reference to the fact that the precautionary principle is:

> based on the understanding that definite scientific proof about some cause and effect mechanisms in the ecosystem may not soon or never be attained. Uncertainty should therefore not be used as a reason for postponing measures to prevent possible environmental degradation.

The Canadian delegation also noted that the principle is:

> based on the understanding that such definite proof may not soon or never be attained.

2.4.3a. *A Shifting Burden of Proof*

An underlying commonality of all the examples of precautionary measures given here is that of a shifting burden of proof.[63] Traditional legal standards in the environmental area have tended to privilege parties accused of degrading the environment; until "proven wrong" such parties can continue the activity in question. The precautionary principle advances an opposite standard: until "proven right," proto-polluters should not be legally permitted to act in ways that carry a non-negligible environmental risk of environmental degradation.

Discussions of this matter sometimes fail to recognize that the "burden of proof" can be understood in one (or both) of two ways. As noted above under Sections 1.4.1 and 1.4.2, the criteria for determining what risks are non-negligible and what regulations are justified can be eased in such a way as to make more risks non-negligible and more regulation justified. Completely certain scientific evidence of harmfulness, for instance, might no longer be required to classify an act as risky and justify the imposition of a certain regulatory standard; this is the first sense in which a burden of proof might be understood to have shifted, as a result of the precautionary principle. We would suggest that it is in this sense that most international legal fora have recognized the need for an altered burden of proof.

However, the burden of proof can shift in a different fashion, namely as a result of a change in identity of the party obligated to prove that a burden either has or has not been satisfied. This is not the same issue as the identity of the parties determining the general evidentiary burden to be satisfied (an issue addressed in Section 1.4.3), but one of the identity of those determining whether in a given case that burden has or has not been met. This is the shift that is contained in most of the indirect precautionary measures listed below, but with the exception of the NAP and the Oslo Commission's Prior Justification Procedure,[64] it is not one explicitly recognized in international legal fora. For effective procedural implementation of the principle, this lack of recognition should be addressed. As Olson has argued, the traditional Anglo-American concept of the burden of proof has been misapplied in environmental cases. Since the burden of proof should rest with the part seeking to change the status quo, then that party seeking to maintain a less-polluted state ("the state before alteration by human activity") should not, as is currently the case, be obliged to carry the burden of proof.[65]

3. The Precautionary Principle in International Law

This section reviews the traditional sources of international law and the particular approach of communicative prescription, and argues for the inclusion of the principle into the body of international law.

3.1. *Traditional Sources*

3.1a. *Treaties*

The first mode of determination is treaty-based. Article 38 of the Statute of the International Court of Justice cites as a source of international law, "international conventions, whether general or particular, establishing rules expressly recognized by the contesting states." Although elements of international environmental treaties have contained a variety of precautionary measures, explicit statements of the principle, as a principle, have not typically been featured in formal treaties. However, a number of recent cases suggest that the precautionary principle may soon feature as a part of treaty-based international law.

The most important example in this regard is unquestionably the revisions to the Treaty of Rome agreed at Maastricht in 1992.[66] The Single European Act of 1987, which placed Community environmental policy on a treaty footing for the first time, stipulated that environmental policy "shall be based on the principles that preventive action should be taken, that environmental damage should as a priority be rectified at source and that the polluter should pay."[67] The Treaty of Rome, as revised at Maastricht, repeats this formula but conspicuously precedes it with the statement that Community policy on the environment "shall be based on the precautionary principle and one the principles that preventive action . . ." etc.[68]

Two points are worth stressing. First, the language employed is of a "harder", binding type rather than a "softer", recommending type. Where Article 130r of the SEA stated principles of the latter type ("preventive action should be taken", "polluter should pay"), the Maastricht revision states simply that Community policy shall follow the precautionary principle. Second, by the inclusion of the conjunctive "and" the precautionary principle is explicitly separated from the previous treaty norms of environmental policy. Although otherwise undefined, the principle is clearly not to be interpreted as new terminology for pre-existing regulatory standards. Moreover, the very lack of definitional elaboration suggests, albeit tentatively, that the principle is, contrary to Handl's (pre-Maastricht) assertion, now a "term of art."

The Oslo and Paris Commissions agreed at their 1990 Annual Meetings to establish Ad Hoc Working Groups on the Revision of the Conventions (OSPARREV) to consider revisions to the Paris and Oslo Conventions in light of, inter alia, the precautionary principle. Resultant revisions are due to be considered at the ministerial meeting of the two Commissions, to be held in France in late 1992. Early indications are that the two Conventions will be replaced by a single Framework Convention to be based, at least in part, on the precautionary principle.[69]

Finally, it should be noted that a certain number of conventions contain precautionary measures or procedures, although they do so without the support of a complementary general statement of the principle. The Convention on Regulation of Antarctic Mineral Resources is probably the most important example in this regard.[70] As Section 3 below demonstrates, there exists a wide range of procedural measures that manifest the precautionary principle; some of these, especially provisions relating to environmental impact assessment, are of a legally-binding form.[71]

3.1b. *Customary International Law*

Drawing partly from the above account of treaty developments, it may also be said that the evidence for the precautionary principle constituting a customary norm of international law is increasingly conclusive. Article 38 of the International Court of Justice cites "international custom, as evidence of a general practice accepted as law." Customary norms have been deemed by the Court to arise when state practice is extensive and virtually uniform, and regarded by states as legally obligatory.[72]

Consider, simply, the importance of the UN Conference on Environment and Development in contributing evidence in support of the emergence of a customary rule of law with universal application. The vast majority of world's leaders were represented at UNCED, the political declarations were therefore authorised at the highest possible level by the largest collection of leaders ever assembled, and the Conventions were, at date of writing, singed by some 157 States.

Even the brief survey of state practice noted in Section 1 above provides considerable evidence of the extension and uniformity of the precautionary principle. It should be stressed that this review is by no means comprehensive; precautionary measures are now contained, both explicitly and implicitly, in numerous interstate conventions, ministerial declarations, and reports and recommendations issued by international organizations, both governmental and otherwise.[73]

Opinio juris may be established in a variety of ways. Recent treaty developments provide prima facie evidence of an emergent sense of legal obligation to adhere to the precautionary principle. Significantly, those parties opposing the formal inclusion of the principle into the terms of the conventions noted above, have argued against that inclusion on grounds of redundancy. For instance, the United States submission to the London Dumping Convention's working group on the precautionary principle stated that: "It is largely because the Convention employs such an approach that it is unclear how the express adoption of a precautionary approach would change decision-making under the LDC."[74] The terms of opposition to the principle, in other words, can demonstrate the prevailing sense of obligation towards it.

State practice may itself constitute evidence of opinion juris. This is especially so in newly-developing areas of law; as states continue to endorse the precautionary principle, customary law can form instantaneously precisely because there is nor pre-existing relevant body of customary international law to be displaced.[75] An informative precedent is provided by the development of rules governing space sovereignty.[76]

During the International Geophysical Year (1957–1958), the urgent need for internationally-collaborative research efforts became widely-recognized. Expectations that the scientific community should not miss out on the chance for extended scientific investigation of phenomena unique to the period, and the lack of a pre-existing customary law of outer-space, allowed for the instant development of a new body of international customary law. The rule that national sovereignty over air space does not extend as far as the minimum altitude required for planetary orbit, grew directly from the cooperative endeavour of the IGY. The analogy with current international environmental developments is obvious enough; the precautionary principle may be seen as emerging in response to uniform expectations surrounding the need for international collaboration, expectations generated by a rapidly growing awareness of environmental degradation and its transnational genesis.

Furthermore, the instantaneous nature of these developments may be exaggerated; norms of precaution are evident well before the principle's international articulation in the late 1980s. As early as 1972, Kirgis argues that in a transnational context United States practice, "after having indicated some reluctance to consider unproven damage, has now evolved into a rather clear recognition of a duty to take account of it, and even to abstain from proposed action if the damage could be serious and if available safeguards do not give substantial assurance of safety." Moreover, the example of United States practice has resulted in "an increasing level of international expectation regarding precaution."[77]

Finally, it may be noted that commentators are coming to a gradual acceptance of the precautionary principle as an element of customary international law.[78] Freestone, for instance, asks:

> Would it be reasonable today (in the current state of law) to presume that a state which has participated in the endorsement of the precautionary principle in a particular sector would be held liable in the future for causing harm (whether to neighbours or commons) for activities in that sector which today are strongly suspected (but not proven) to cause substantial harm?"[79]

His assessment is that "the evidence suggests that the answer may well be affirmative." He concludes:

> At present this is probably arguable on a sector by sector basis, but if present trends continue this may be a general requirement of environmental law in the not too distant future.[80]

De La Fayette commenting on the Second North Sea Conference, has observed that:

> it [the precautionary principle] has become the fundamental norm underlying the practical measures taken then and at the Third Conference. In adopting the precautionary principle, States have recognized that it is necessary to take action to reduce pollution in the sea even before there is positive scientific evidence that damage has occurred or will occur . . . In recent years, this approach has become crucial to actions now being taken or proposed in other aspects of environmental protection, such as global warming and the destruction of the ozone layer.[81]

3.2. *Functional Determinations of International Law*

3.2a. *Law As Communicated Prescription*

A final mode of determination is provided by the functional communications model of international law, associated with Lasswell, McDougal and Reisman.[82] In this model, international law-making is regarded as a form of communicated prescription involving "the mediation of subjectivities from a communicator to an audience and, in successful cases, a reception and incorporation by the intended audience, resulting in a set of appropriate expectations that are supposed to influence behaviour and, contingently, to alert community enforcement responses when deviations are deemed to threaten public order."[83] Specifically, prescriptive communications must encompass a policy content, an authority signal, and an intention of control.[84] These criteria are increasingly satisfied by the precautionary principle. The policy content of the principle is substantial and specific, as Section 3 will demonstrate.

The authoritative and controlling aspects of prescriptive communications are more complex variables. Reisman notes that it is the community of communicates, rather than communicators, that confers the necessary authoritative status on prescription, and this community has expanded in tandem with contemporary "extraordinary increases in the variety, number, and range of channels in international political communication."[85] Prescriptive expectation is also sustained by a control element, the signalled capacity and willingness to enforce the norm in question by means of political power. The precise combination of authority and control communications required to effect law will vary on a case by case basis. Authority will play a greater role when the community of respondents has a relatively developed sense of collective identity and common interest; where this same sense is relatively underdeveloped, control signals are correspondingly more significant.[86]

We contend that the authority and control elements of precautionary principle communications are by now sufficient to render those communications prescriptive, and thus constitute a part of international law. Certainly, it is the case that, whether or not that sufficiency can be conclusively established or not, the trends towards sufficiency are clear enough.

3.2b. *The Communication of Authority*

First, the precautionary principle has gained prominence in international environmental fora precisely because of the growing recognition that for a variety of reasons pre-existing regulatory standards lacked authority. The precautionary principle is a response to the perceived inadequacies of predictive science to guide law and policy under conditions of acute environmental interdependence and a perception of threat of irreversible harm to the biosphere. Awareness of such scientific inadequacies and environmental interdependence has grown among both policy-makers and broader publics. That interdependence has increasingly taken the form of a transnational – and even transgenerational – recognition of a common human identity as dependents of the global environment with a demonstrable collective interest in its protection. It other words, the forces responsible for the emergence and development of the principle themselves suggest that it is received as highly authoritative.[87]

Second, it might be argued in support of this first point that those challenging the authority of the precautionary principle now do so in ways that indicate the extent of that same authority. American objections to the principle have moved from directly-stated outright opposition,[88] through efforts to redirect the precautionary focus in ways incompatible with the precautionary principle,[89] to claiming that current implementation of the principle is so advanced as to make explicitly-agreed statements of the principle unnecessary.[90] Acceptable or legitimate terms of opposition to the principle have changed as the authority of its communication has increased.

3.2c. *The Communication of Control*

Third, current developments indicate that the principle will be increasingly communicated in conjunction with detailed articulations of control intent.

The 1992 U.N. Framework Convention on Climate Change provides for the Conference of the Parties (C.O.P.) to be the authority exercising control over the legal commitments made by the states. The C.O.P. will meet on a regular basis. Questions regarding interpretation and implementation, for instance, will be brought by Parties or by other subsidiary institutional organs (such as the permanent Susiduary Body on Implementation) to the Conference of the Parties which will then institute consultative proceedings by means of an ad hoc panel. The panel will report within a year; the report will be considered by the Conference which will then issue a decision, by a two thirds majority if necessary. Detailed proposals for dispute settlement procedures as well as reporting requirements are also elaborated in the agreed text which contains, what is for a Framework Convention, an unsurpassed degree of detail concerning institutions and institutional procedures.[91]

A similarly high degree of control intent was evinced by the U.N. Security Council in passing Resolution 687, thereby reaffirming Iraq's liability for any environmental damage and depletion of natural resources caused by the invasion and occupation of Kuwait, and created a fund to meet claims for compensation to be administered by a UN-established Commission. This may be the most substantive form of control effected thus far in international environmental law at the global level.[92]

At the regional level, European Community and Nordic environmental law very evidently communicate control intentions.[93] The Community's principle of direct effect, for instance, asserts the primacy of Community law over national law; European precautionary environmental directives are thus enforceable through national courts.[94]

Additionally, there have been some suggestions, notably form the European Parliament, that the nascent European Environment Agency should be granted enforcement powers analogous to those of the Environmental Protection Agency in the United States.[96] See also S.4.2.3c "Environmental empowerment".

3.2d. *Combined Communications of Authority and Control*

Control intent may become increasingly evident at the level of monitoring and reporting regimes. Recent examples, such as the European Environment Information and Observation Network,[97] and the London amendments to the Montreal Protocol, suggest that such regimes may be becoming more stringent and intrusive. Control at this level can establish whether agreed policies and laws are being followed in the prescribed fashion. This exposure of adherence or defection from the terms of an international environmental regime can combine with signalled authority to raise the prescriptive content of a communication to the level of law-making.

The proposed Implementation Committee of the Montreal Protocol, for instance, is essentially a very public complaints procedure; it is not designed to control implementation of the Protocol directly, but to effect transparency over issues of compliance or non-compliance.[98] Once a party's adherence to or violation of the Protocol is so visibly demonstrated, the authority content of the prescription can then come to bear. The Climate Convention requires that each Party communicate to the Conference of the Parties through the Secretariat, *inter alia*:

(a) A national inventory of anthropogenic emissions by sources and removals by sinks of all greenhouse gases not controlled by the Montreal Protocol . . .

(b) A general description of steps taken or envisaged by the Party to implement the convention;[99]

Such reports obviously carry a potential for exposing non-compliance with the terms of the Convention, and thus for allowing more effectively authoritative prescription.

4. The Behavioral Impact of the Precautionary Principle

Thus far, we have attempted to alleviate the definitional ambiguities sometimes attributed to the principle, and to demonstrate the status of the principle in international law. In this section, we seek to answer the question of whether the precautionary principle matters. Can – or does – the principle impact on environmental behaviours? Is it effective? These questions can be answered at two broadly delineated levels.

Since, we are concerned with the development of the principle within international environmental regimes, the principle will take effect only inasmuch as those regimes are effective. This moves the analysis towards one of the most heated controversies raging in the current international relations literature, namely the issue of whether international regimes themselves matter in terms of influencing behaviour.[100] This paper has no ambitions to resolve this dispute; however, it will argue that there are specific aspects of the precautionary principle suggesting that regimes having recourse to its prescriptions at least partially overcome some of the arguments advanced by those asserting the general causal irrelevance of regimes.[101] At a general level, the precautionary principle functions well in overcoming tendencies towards international non-cooperation. Reasons for this successful functioning are given below in Section 3.1.

The second level of debate centres around the capacity of the principle, once accepted at a general level, to guide the formulation and implementation of specific environmental procedures. The principle, we suggest, should be recognized as functioning effectively in determining particular policy instruments. The reasoning behind this contention, as well as some of the procedures involved, are given below in Section 3.2.

4.1. *The Precautionary Principle as a General Cultural Standard of International Environmental Cooperation*

4.1.1. *Meaningless Cooperation? A Realist View of International Regimes*

Many commentators, particularly from the realist or neo-realist school of international relations, would deny the possibility – both theoretic and actual – of the principle's exerting any influence on international behaviours.

Realists assume that such rules as are agreed at the international level will not be in and of themselves causes of cooperation or noncooperation among actors on the international level. International rule-making and rule-following are determined by questions of state power and state interest, both of which are exogenously given, that is uninfluenced by the formulation and application of rules. International environmental regimes, like all regimes, are nothing more than a liberal institutionalist fig leaf on the anarchical body (real)politic. A proposed standard of international behaviour, like the precautionary principle becomes a case of "mere words" since it serves merely to codify existing constellations of power and interest. It is itself causally irrelevant.[102]

In the conventional realist paradigm of international relations, this is taken to be the case whether the principle codifies mutual or divergent interests. Codification of mutual interests puts an apparently cooperative stamp on a series of behaviours that would have been pursued even in the absence of such "cooperation" by states seeking to maximize the realization of their interests.[103] Where state interests are divergent rather than mutual, international agreements will only ever constitute agreements to disagree. [104] Rules will be agreed in language so vague and open to interpretation, that states will be able to comply with such rules while advancing their divergent self-interests. Thus Morgenthau claims that states can and will:

> naturally interpret and apply the provisions of international law in the light of their particular and divergent conceptions of the national interest. They will naturally marshall them to the support of their particular international policies and will thus destroy whatever restraining power, applicable to all, these rules of international laws, despite their vagueness and ambiguity, might have possessed.[105]

The Noordwijk declaration on climate change, for instance, explicitly agreed on the necessity for the stabilization of greenhouse gas emissions "as soon as possible."[106] Here, each signatory can interpret "possible" according to its own self interest. The apparent U.S. interpretation was that such stabilization would not be possible until the economic implications of stabilization were better understood, and the scientific uncertainties surrounding climate change were reduced.[107]

Does the precautionary principle fit either of these two patterns of agreement? Is it, in other words, purely epiphenomenological?

4.1.1a. *Bergen: An Empirical Counter-example*

The Bergen Conference strongly suggests that the terms of the precautionary principle cannot in practical terms be considered open-ended.[108] The draft ministerial declaration for the Conference included a paragraph referring to the principle which stated that:

> . . . environmental measures must . . . anticipate, prevent, and attack the causes of environmental degradation, even if final scientific proof is lacking. Doubt should not be used as a reason for postponing measures to prevent environmental degradation.[109]

During the month prior to the Conference, the U.S. strenuously sought to block the inclusion of this language. A State Department telex laid out the basis of the American objections: that there was no-internationally agreed definition of the principle, that the declaration language above was too open-ended, and that the adoption of the precautionary principle would impose a "single-criterion" decision rule on complex issues. The telex apparently stated that:

> The U.S. government strongly believes that we cannot commit ourselves to "anticipate, prevent and attack the causes of environmental degradation, even if final scientific proof is lacking" . . . (and) does not believe that we can lightly accept above language in the expectation that it will be forgotten with the passage of time.[110]

The tension in this communication is obvious enough: the principle is deemed to be simultaneously too broad to be meaningful and too constraining of behaviour. In opposing the principle, the U.S. subsequently found itself embarrassingly isolated, losing in particular the support of two previous allies on issues of greenhouse gas reduction, Canada and the U.K. This position of isolation was hardly likely to have been accepted on the grounds that the principle was too open-ended. Certainly the realist paradigm could hardly argues this line, since meaningless agreements do nothing to prohibit the advancement of state interests, and will therefore not be opposed by states seeking to realize those interests. More plausibly, the U.S. did regard the inclusion of the precautionary principle as imposing a constraint on the successful realization of American national interests. This interpretation is given added force by the concurrent U.S. objection to the inclusion of a statement calling for immediate adoption of CO_2 emission reductions, namely that such targets would prejudge the outcome of the IPCC process. Widespread endorsement of the precautionary principle, however generally expressed, would obviously curtail this particular line of objection.

There are two points to stress here. First, whatever its specific procedural indeterminacies, the precautionary principle does have interpretive limits. The earlier Noordwijk commitment to postpone actual reduction targets until the IPCC issued conclusions and recommendations was obviously threatened by the precautionary principle. It is therefore meaningfully prescriptive rather than meaninglessly open-ended. Second, states are prepared to incur isolation in international fora as a result of their opposition to the principle. It is hard to imagine why such opposition could be thought worthwhile, if the principle was assumed to be entirely epiphenomenological.

International agreement of the precautionary principle does appear to be regarded as meaningful by parties to that agreement; it does therefore amount to a substantive rule or norm. Parties to the principle have agreed to behave in a meaningfully cooperative fashion.

4.1.2. *Unattainable Cooperation? Dealing with Interdependence*

That states attribute normative force to the precautionary principle does not of course logically entail that they will then in practice follow its prescriptions in a cooperative fashion. The reasons for this are well-known and require only a brief rehearsal here.[111]

International cooperation on matters of environmental degradation constitutes a collective action problem. The state of environmental interdependence is such that cooperation is the only means towards either preventing further environmental damage or rectifying existing damage; unilateral or uncoordinated endeavour will have only a negligible impact.[112] But although the need for multilateral coordination may therefore be recognized, that coordination may not be forthcoming. This is because parties to a precautionary agreement may calculate that their own defection from that agreement would not deny them the benefits of multilateral environmental coordination, while at the same relieving them of any of the costs of such coordination. At the same time, states are aware that other states may be going to make this calculation, and therefore conclude that environmental cooperation is doomed from the outset. Non-cooperation, for one reason or another, therefore results, despite the fact that states may recognize that cooperation is in every parties' interests.

A growing awareness of transnational environmental interdependence, does not guarantee international environmental cooperation. The demand for cooperation (in the form of the precautionary principle, for example) does not necessarily mean that the precautionary principle will be effectively supplied as a solution to the problem of environmental degradation. To suppose that it does is to commit what has been termed the "functionalist fallacy", analogous to supposing that because the prisoners of the famous prisoners' dilemma recognize their dilemma and its interdependent nature, they will therefore simply resolve it.[113]

But there are peculiarities of the precautionary principle as an environmental standard that mitigate the force of this argument.

4.1.2a. *The Impact of Uncertainty.* The precautionary principle advances environmental regulation on the basis of uncertainty; uncertainty is premised as a sufficient justification for regulatory measures. Cause and effect relationships are so inadequately understood, and the potential costs of regulatory inaction so high, that regulatory inaction is unjustifiable. The principle therefore argues for worst case planning. As a result, cooperative endeavour are that much likelier to be entered into; the risk of falling foul of the prisoners' dilemma is diminished by the costs of failing to cooperate in the face of uncertainty. The risks and costs of possible defections from agreed standards are, like the risks and costs involved in the imposition of regulatory mechanisms, taken as relatively insignificant next to the risks and costs of failing to agree such standards in the first place.

The key point to stress here is that the precautionary principle, as a standard of international environmental cooperation, is relatively self-fulfilling. Once the uncertain nature of possible non-cooperative outcomes is recognized (itself a result of the principle being publicly and repeatedly expounded at the international level), the conditions are in place for a cooperative solution to the dilemma of collectively-induced environmental degradation. In game theoretic terms, the payoffs resulting from non-cooperation are so uncertain and potentially damaging, while the payoffs from cooperation are so high as a result of addressing those uncertainties, that prisoner dilemma-type situations no longer aries. The incentive to defect unilaterally from a cooperative arrangement (or free ride) is outweighed by the risk of cooperation breaking down as a consequence of that defection (possibly in combination with other defections).

Cooperative arrangements in the security area provide an informative parallel. In discussing the prospects for a substantive climate convention, Skolnikoff has despondently noted that:

> The central problem is that outside the security sector, policy processes confronting issues of uncertainty do not normally yield policy that has high economic or political costs.[114]

Obviously, what is distinctive about the security sector is the perceived potential for very high but uncertain costs accruing as a result of a failure to take sufficiently stringent security measures.[115] Clearly if environmental cooperation is seen to address a similar type of problem, and environmental non-cooperation to threaten a similar set of costs, then the environmental collective action problem will be similarly resolvable. *The precautionary principle stresses that this is precisely how environmental issues must be addressed. Inasmuch therefore as it becomes demanded at the international level, the precautionary principle goes a long way towards supplying the conditions necessary for its realization.*

Moreover, the environmental uncertainties that the precautionary principle stresses are unlikely to be resolvable within the short term; the complexities of environmental interdependence, especially in the context of cumulative forms of pollution, make this near-certain. Thus the demand for regulatory standards in the face of uncertainty will persist, and furthermore the likelihood of this persistence is known currently. Parties to agreements therefore have a greater incentive to keep them; the shadow of the future is sufficiently long to make the incentive for defection untempting. The possibilities for persistent cooperative benefits, nurtured by a developing tradition of cooperation, are unlikely to be rejected for the sake of what may turn out to be a one-time pay-off. As the game theory literature now stresses, the prospect of continued reiterations of the prisoners' dilemma stretching away into the future for an unspecifiable but lengthy period of time, is likely to induce players to develop strategies for the realization of cooperative equilibria on a long-term basis.[116]

4.1.2b. *Transnational Identities*. The precautionary principle has partly grown out of, and certainly contributes to, the development of transnational identities. At the environmental level, interdependencies are increasingly regarded as so far-reaching that they have taken on an intangible form. Individuals and groups are increasingly concerned with environmental impacts that may not harm them directly, but because they effect some aspect of global environmental well-being, are deemed to violate their environmental interests. The notion of benefiting environmentally at someone else's expense is increasingly viewed as anachronistic. States, groups and individuals are concerned to protect the environment not just for the tangible and direct benefits that they receive as a consequence, but because that protection, wherever it takes effect, is regarded as an end in itself. The individual's environmental identity is increasingly with their environmental fellow travellers, the community of individuals both now and in the future.

Transnational identities can have a substantial impact on a collective action dilemma of an international nature. The benefits of free-riding on the cooperative arrangements of others simply dissipates as the costs of "not doing one's part" rise. If individuals are as concerned with the environmental health of other individuals as with their own, then nothing is gained – even in the short term – by defecting from established environmental regimes. The precautionary principle can thus become a fixed standard analogous to the white line down the middle of a road. The rule that the white line embodies does constrain behaviour, but in an entirely innocuous fashion, since no party has an interest in departing from the norm; to do so is as potentially damaging for them as it is for any other road-user. Collective action problems need no longer arise.

4.1.3. *Towards Genuine International Law?*

Moreover transnational identities point towards the development of a body of law that addresses problems of global interdependence directly, rather than indirectly through the medium of state sovereignty. International environmental law has already partially dispensed with that medium; the expansive role of transnational epistemic communities in the construction and implementation of particular regional environmental regimes had been convincingly demonstrated.[117] These closely self-identifying groups of policy-makers and scientists have acted directly in response to their perception of a transnational environmental interest, and thus begun a conceptual redefinition of sovereignty.

Because the precautionary principle arises from, and reciprocally stimulates, global epistemic communities, it constitutes a move towards a similar redefinition at the global level. Inasmuch as it represents a growing transnational interest in the protection of the environment as an end in itself, it serves the development of the law of an international society, the society of all societies, and no longer the international law of the society of independent sovereign states. Allott has stressed that it is a particular

> consciousness which makes possible, which legitimates, which naturalizes, the way in which we conceive of international society and international law.[118]

The precautionary principle represents the beginnings of a change in this consciousness; the extent of the uncertainties it seeks to resolve, and the transnational identities consequently generated, threaten the adequacy of international law and policy developed on the basis of existing conceptions of sovereignty. Hence a reconceptualization of international society and law becomes possible.[119] These reconceptualizations are obviously only at an incipient stage, but they hold out the promise of international law that will function as the law of an actual international society: one that is not conceived of as a community of states with separate interests, but one that is increasingly and transnationally concerned with the protection of its entire environment.

4.1.4. *Transaction Cost Arguments*

One final advantage of the precautionary principle is its adaptability to a conspicuous trend in international regime development, particularly in the environmental area. *Ad hoc* approaches to international cooperation have become increasingly replaced by very broad forms of institutionalized multilateralism. Framework conventions, with only general norms, are now agreed prior to the regime's specific cooperative elements; these tend to be contained in protocols or annexes, which are themselves subject to regular, and relatively simplified review, procedures.

This flexible framework is designed to reduce the transaction costs of international cooperation; not every separate cooperative effort requires "the elaborate and time consuming exercises in diplomacy" required by the traditional ad hoc approach. Moreover, with a broad framework in place, parties will have sufficiently high sunk costs in cooperative outcomes to reduce the incentives for "opportunistic behaviour."[120] The empirical record, as Thacher has noted, suggests that the more successful, and more rapidly concluded environmental agreements have tended to follow this path.[121]

As noted above, the precautionary principle can be stated in general abstract terms that are both meaningful and effective. Yet as the following sections demonstrate, the principle is also capable of guiding specific environmental regulatory procedures. This dualist nature obviously recommends itself to the dualist nature of emergent environmental agreements noted in this section.

4.2. *Procedures*

4.2.1. *Operationalizing Precaution*

The question remains as to whether the precautionary principle can be procedurally operationalized, even once it has been accepted as the basis for general environmental law and policy. Those opposing the principle tend to seize on its supposed procedural indeterminacy. The Group of Experts on the Scientific Aspects of Marine Pollution, for instance, have claimed that:

> It is clear that a single instrument, such as the precautionary principle, cannot deal adequately with the complexities of environmental management human development.[122]

Greenpeace International has rejoindered:

> The precautionary principle is not a single policy instrument, but a general principle applicable to all policy instruments.[123]

We agree. In essence, any environmental regulatory procedure will be precautionary if it reduces the scope and magnitude of environmental impacts whose effects are uncertain, but which are deemed to carry a non-negligible environmental risk. This provides two general rules of operationalization.

A. Regulatory standards will tend to be more precautionary, the closer their point of application to the point at which such impacts are generated.
B. For any given form of environmental regulation, whatever its point of application, there exists a spectrum of possible degrees of precaution.

Rule A follows from the inherent uncertainties of predictive science, discussed above in Section 1.4.3. Since our knowledge of environmental cause-and-effect relationships is inherently limited by such uncertainties, the more regulatory standards act to reduce the range and abundance of possible causes of environmental degradation, the greater the reduction of uncertainty, and the more effective the application of the precautionary principle. Thus standards that apply to the generation of environmental inputs, will tend to be more precautionary than those which apply to those inputs once generated.

In the case of industrial waste, for instance, a standard that takes effect before the generation of waste occurs will be more precautionary than any control on the disposal of waste once generated. It is important to note that this is not because such standards will be more preventive, although the presumption is that will indeed be the case, but because waste reduction at source eliminates or reduces the generation of environmental inputs whose effects remain uncertain. There may be many arguments for reductions at source, such as economic efficiency or the elimination of known pollutants, but the precautionary principle constructs its hierarchy of waste management options on the basis of uncertainty reduction.[124]

Rule B is straightforward enough. Any given regulatory standard advanced on a precautionary basis can be more or less precautionary. End-of-pipe emission standards while not as precautionary as various forms of clean production are not uniformly precautionary; if they are imposed in order to reduce emissions of inputs with uncertain environmental effects, then the more they effect such reductions the more precautionary they will be. If they are not imposed on that basis, but rather with the intention of reducing inputs with more certain environmental effects, then the precautionary principle is not involved. *Thus, whether or not a given type of control (such as an emission standard) is precautionary or not, will depend upon the context of its application.* In this sense then, Gundling is right to emphasize that "it is important not to identify the principle of precautionary action with a specific regulatory approach."[125] A number of international agreements of one kind or another adopt or endorse regulatory measures "further to" or "taking into consideration" the precautionary principle; this statement of avowed intent is not enough to guarantee that such measures will in essence be precautionary.[126] Precautionary discourse may be employed purely to legitimate the imposition of what are exclusively preventive standards.

4.2.2. *A Typology of Precautionary Procedures*

There are essentially two types of precautionary measure: those that create conditions likely to result in de facto regulation and those that impose de jure precautionary standards. The distinction between these two procedural types is not absolute, but is, we suggest, conceptually useful. The key distinction is that procedures of the first type are prescriptively more indirect than those of the second type; they create a legal or political environment of incentives and disincentives which will tend to generate behavioral adherence to the precautionary principle. Procedures of the second type, on the other hand, either directly require behaviours that adhere to the principle, or prohibit those that do not so adhere, or both. Some procedures amalgamate both types into a single form of regulation.

4.2.3. *Indirect Precaution*

There are a variety of indirect precautionary measures: strict liability statutes, insurance mechanisms, forms of "environmental empowerment" and environmental impact assessment comprise an incomplete list.

4.2.3a. *Strict Liability*. Strict liability statutes are precautionary, since they act as an incentive against generating environmental inputs that may subsequently have a degrading effect. They stimulate proto-polluters to assess the likely effects of their actions before they take place, and in circumstances where those effects remain uncertain they create a strong incentive to refrain from the potentially damaging act altogether. Already a feature of some domestic legislation, especially in North America,[127] provisions for strict liability are now gaining greater prominence at the international level also.

An amended proposal for a Council Directive on Civil Liability for Damage Caused by Waste states in Article 3 that:

> The producer of waste shall be liable under civil law for the damage and injury to the environment caused by the waste, irrespective of fault on his part.[128]

The preamble of the same directive notes that:

> . . . in view of the risk inherent in the very existence of waste, the strict liability of the producer constitutes the best solution to the problem.

Moreover, the draft directive provides for joint and several liability. This will increase precautionary behaviour still further, since proto-polluters cannot hope to exploit uncertainty about the precise causal responsibility for environmental damage in order to escape lability.

If recent trends are indicative, strict liability is a precautionary standard that is likely to become well entrenched in the international normative milieu.[129]

4.2.3b. *Insurance Mechanisms*. As noted above in Section 1, the precautionary principle is closely tied to notions of insurance. Environmental insurance schemes constitute perhaps the clearest operationalization of the principle. like strict lability statutes, insurance schemes are precautionary, since they too create substantial financial incentives against generating environmental inputs that may subsequently have a degrading effect, but which cannot currently be conclusively shown to do so. This, of course, is only the case where the insureds are also the parties – or perceived to be the parties – whose actions are assumed to be potentially, although uncertainly, responsible for the environmental damage that has been insured against.[130] Furthermore, if successfully implemented they could come to influence environmental liability regimes, since they implicitly strengthen the legal claims of environmentally-damaged parties.

The 1971 International Convention on the establishment of an International Fund for Compensation for Oil Pollution Damage provides, under ceratin restricted circumstances, for the payment of compensation to any person suffering oil pollution damage. It is funded by those potentially responsible for oil pollution damage, namely the oil industry, via levies on oil importers.[131] The 1960 Paris Convention on Third Party Liability in the Field of Nuclear Energy, together with the 1963 Brussels Supplementary Convention and subsequent protocols, also effectively establishes an international insurance pool; it is funded by states on a basis which takes into account gross national product and unclear energy capacity.[132]

4.2.3c. "*Environmental Empowerment*".[133] As detailed in Section 1.4.3 above, the precautionary principle recognizes the need for publicly-generated political judgements of what counts as the correct regulatory response to uncertain environmental possibilities. The precautionary principle cannot take hold therefore unless regulatory decision-making procedures provide meaningful points of access for these subjective determinations. Environmental empowerment serves as a useful blanket term to cover the provision of these points of access. A variety of measures may be mentioned in this context.

For access to be meaningful, environmental information must be made available to interested parties. A wide range of international fora have called for provision of greater public access to environmental information, and for a recognition of the significant environmental impact that will result from that access. The Bergen Declaration, for instance, explicitly links the realization of sustainable development at global, regional and national levels to the establishment of necessary democratic institutions and policies, including public participation and information.

Significantly, the European Environment Agency in conjunction with the European Environment Information and Observation Network, was established with the express purpose of, inter alia, providing member states with:

> objective, reliable and comparable information at the European level enabling them to take requisite measures to protect the environment, to assess the results of such measures and to ensure that the public is properly informed about the state of the environment . . .[134]

Additionally, the Agency will be subject to the terms of the 1990 European Council Directive on Freedom of Access to Information on the Environment.[135] Building on the Fourth EEC Environmental Action Programme (which called for devising "ways of improving public access to information held by environmental authorities"), it requires, albeit with so major caveats, that member states should:

> ensure that public authorities are required to make available information relating to the environment to any natural or legal person at his request and without his having to prove an interest.

The Directive's recital states explicitly that:

. . . access to information on the environment held by public authorities will improve environmental protection.

It has been noted that since the Agency is open to countries which are not members of the European Community (Article 19) it may emerge as a model for environmental monitoring arrangements at the international level.[136] Various Central and Eastern European countries have indicated their intention of participating in the Agency's work.[137]

Once information is available, efforts must be made to ensure that such information can be brought to bear by interested parties, such as international non-governmental organizations, both on decision-making within international agreements and on implementation of agreed standards. At the decision-making level, NGO participation allows environmental uncertainties to be publicized and the subjective assessment of the correct response to those uncertainties democratically widened. With the notable exception of the 1974 Nordic Convention, the international legal standing of such parties has been distinctly inadequate, although there has been some recent progress in this regard. The U.N. Framework Convention on Climate Change provides that:

any body or agency, whether national or international, governmental or non governmental, and which has informed the secretariat of its wish to be represented at a session of the Conference of the Parties may be so admitted unless at least one third of the Parties object.

NGOs "having special competence in the field" may be granted a consultative status at the International Atomic Energy Agency, and observer status may be accredited to NGOs "concerned with developing uses of unclear energy for peaceful purposes."[138]

From the late 1980s, NGOs have been able to assume a more substantial role in the enforcement of EC environmental law; they can, for instance intervene as interested parties in certain cases before the European Court of Justice in which environmental issues are at stake. The UN/ECE Guidelines on Responsibility and Liability regarding Transboundary Water Pollution allow non-profit making organizations competent in environmental matters:

. . . the right to institute proceedings or to resort to other administrative relief in the case of transboundary water pollution."[139]

4.2.3d. *Environmental Impact Assessment.* The UNEP Governing Council had defined environmental impact assessment to mean:

. . . an examination, analysis and assessment of planned activities with a view to ensuring environmentally sound and sustainable development. EIA goals and principles . . . are necessarily general in nature and may be further refined when fulfilling EIA tasks at the national, regional and international levels.[140]

Obviously all forms of EIA are directly precautionary in at least a minimal sense; if a particular process is required to submit to an EIA, then that requirement alone constitutes a form of regulation predicated on addressing uncertainty about the effects of future environmental inputs. An EIA may be required before any environmental inputs are generated, or it may be applied to the dispersement of such inputs into the environment, or it may be applied to some intermediate stage, such as the recycling of such inputs into the production process. All these forms of regulation are direct, but only precautionary to a limited extent. That extent is determined by two factors: first, the criteria a given process must satisfy to trigger the requirement of an EIA, and second, the stringency of the EIA process itself.

The current norm is the "triggering"-criteria to be listed positively in the form of grey lists of substances and processes (often attached as annexes or appendices to the Convention in question), that because they meet certain quantitative or qualitative criteria of (typically) toxicity, bio-accumulability or persistence, trigger EIA requirements. Unless banned completely (or "blacklisted"), all other substances and processes are not subject to the regulatory standard. Obviously, these listings can be assembled according to more or less precautionary criteria. Recently, suggestions have been made for negatively listed criteria. Friends of the Earth International, for example, have argued that the New Assessment Procedure proposed for the London Dumping Convention "should actually be a reverse listing of waste categories that are allowed to be considered for ocean dumping."[141] In other words, all ocean dumping of waste should be banned except for a "white list" of substances and processes that may be allowed assuming that an assessment procedure is satisfied.

In themselves, therefore, EIAs amount to a form of mandated information provision, with precautionary effects similar to those made possible by environmental empowerment; they are precautionary enabling devices. The various Prior Consultation Procedures of the Oslo Commission, for example, require parties to the Oslo Convention that propose allowing certain specified activities within their national jurisdiction, to present a report on those activities and their effects, on which other parties may subsequently comment,[142] As Hey has observed, this does not directly prohibit the party in question from ignoring others' comments and interpreting the report in its own fashion.[143] To go beyond a basic information-provision function, criteria must be established for an EIA to satisfy: to be more than precautionary in the limited but direct sense noted here, an EIA must demonstrate that a proposed activity will meet certain precautionary ends.

The criteria that the results of an EIA must satisfy in order to legitimate the proposed activity vary from directly precautionary through indirectly precautionary to unprecautionary. An example of a potentially unprecautionary EIA procedure was that called for by the Dutch delegation to the OECD Environmental Ministers Conference, January, 1991. The delegation proposed that "those who engage in activities which may irreversibly jeopardise the environment (according to early warnings) are asked to submit evidence that their activities are in accordance with the carrying capacity of the environment."[144] It is significant that the burden of proof is on the would be polluter to demonstrate the safety of their proposed activities.[145] Nevertheless, the language of assimilative capacity may raise concerns, if the standards by which carrying capacities are set are not themselves precautionary, then EIA might not merely be unprecautionary but could in point of fact actively legitimate anti-precautionary law and policy.

In most cases the relevant criteria are only indirectly precautionary; they provide a necessary means to the establishment of subsequent and direct precautionary measures. For example, an EIA that was conducted – as UNEP recommends – "with a view to ensuring environmentally sound and sustainable development," would only be directly precautionary to the extent that specific regulatory, precautionary means were established in order to realize "environmentally sound and sustainable development." Similarly, EIAs that are required to show that a proposed activity is "environmentally safe",[146] that it will not adversely affect the . . . environment on a local or regional level,[147] has "no direct effect on human and animal health",[148] that its alternatives are not "practical",[149] that "it would not cause significant adverse effects on global or regional climate or weather patterns" nor "significant changes in atmospheric, terrestrial or marine environments"[150] nor "significant and harmful changes to the Convention area"[151] nor "unacceptable changes" to the environment,[152] that potentially adverse effects are being "minimized",[153] that "should an adverse environmental impact be foreseen, the necessary corrective measures are formulated in the early stages of the proposed project"[154] and so on, will only be precautionary to the extent that criteria for these ends are set in a precautionary fashion.

A lack of specified ends may not always hamper the contribution of an EIA procedure to the application of the precautionary principle. The European Community's 1985 Council Directive on the Assessment of the Effects of Certain Public and Private Projects on the Environment requires that for those projects subject to an EIA, member states must provide a substantial range of information, that includes as one element only:

> A description of the measures envisaged to prevent, reduce and where possible offset any significant adverse effects on the environment.[155]

In this context, EIA is designed largely to provide an informational background against which the need for regulatory measures – including the precautionary – can be assessed and particular policies subsequently adopted.

EIA procedures that specify directly precautionary criteria that a proposed activity must be shown to satisfy, are a relatively recent development. An example is provided by currently-debated changes within, and to, the London Dumping

Convention. Under the current terms of the LDC, certain substances cannot be dumped at sea (except in exceptional cases). These substances are listed in Annex I of the Convention. Annex II lists, although in an indeterminate fashion, a series of substances for which dumping was contingent upon the issue of a nationally-provided permit. National authorities can issue such permits "only after careful consideration of all the factors set forth in Annex III," which include the characteristics and composition of the matter to be dumped, the characteristics of the dumping site and dumping method, and other general considerations such as "the practical availability of alternative land-based methods of treatment, disposal or elimination, or of treatment to render the matter less harmful for dumping at sea."[156]

This is at best only indirectly susceptible of a precautionary interpretation. An Annex II substance can be dumped merely on the basis of a "careful consideration" of a series of characteristics. Those characteristics are relatively open-ended; there are no specified forms, or parameters, of those characteristics that would lead to the proposed dumping being denied a permit. For instance, there are no criteria given for determining the "practical availability of . . . treatment to render the matter less harmful for dumping at sea," nor is there any agreed commitment to do more than carefully consider such practical availability were it to exist.

After the Convention's Tenth Consultative Meeting in 1986, the Scientific Group on Dumping established an ad hoc Group of Experts on the Annexes, which after three meetings, produced a draft New Assessment Procedure (NAP) for consideration at the Fourteenth Consultative Meeting in 1991. The NAP will be further reviewed at the fourth meeting of the Group of Experts (at which several NGOs will be present in a participatory capacity), and then be reconsidered at the Fifteenth Consultative Meeting in 1992.

From the outset, one of the official "guiding principles" of the Group of Experts has been:

> that any new structure for the operational procedure of the Convention should reflect a precautionary approach to dumping.[157]

And the NAP's introduction confirms that:

> In applying the draft NAP . . . it is important that national authorities adopt a precautionary approach to the introduction of substances into the environment and actively pursue measures that will reduce contamination where there is reason to suspect that harmful effects may occur, even though stringent proof of a cause-effect relation may be lacking.[158]

This kind of approach is evident, for instance, in the NAP's provisions for the assessment of alternatives to disposal of waste at sea. In place of a "careful consideration" of whether such alternatives are "practical", there is now a two stage process: first, a waste prevention audit, and second, a comparison of waste management options.[159] The former requires that "applications for permits should be refused and existing permits should be reviewed" if the "feasibility" of each of a number of techniques for waste prevention "have not been adequately addressed." These techniques include product reformulation, clean production technologies, process modification, input substitution, and on-site, closed-loop recycling. Waste management options are considered.

> Applications to dispose of wastes at sea should demonstrate that consideration has been given to each element in a hierarchy of waste management options.[160]

The hierarchy in "an order of increasing environmental impact" includes the waste prevention measures cited in the waste prevention audit above, off-site recycling, re-use, destruction of hazardous constituents, treatment to reduce or remove the hazard, and disposal into land, air and water. The draft concludes:

> Generally speaking, a permit to dispose of wastes at sea should be refused if opportunities exist to recycle, re-use or treat the waste without risk to human health or incurring disproportionate costs. The practical availability of other means of disposal should be considered in the light of a comparative risk assessment involving both sea disposal and the alternatives.[161]

Clearly, the NAP is an EIA of a new order. Proposed activities must demonstrate that they satisfy specific precautionary procedures. The precautionary impact of this type of EIA is thus more direct than indirect. A proto-polluter is required to demonstrate under the terms of the assessment procedure that certain specific regulatory standards are either in place or not needed even under the precautionary framework of the NAP.

4.2.4. *Direct Precaution*

Any regulatory measure will be directly precautionary if it seeks to limit specified environmental impacts whose consequences are uncertain, but which carry a non-negligible risk of being environmentally damaging. Particular activities are either prohibited or legally mandated. Several such measures have been mentioned above in relation to the NAP. The quantified restrictions on CFC production contained in the amended Montreal protocol are another example of directly precautionary measures, as would be any restrictions on CO_2 emissions authorised by the Climate Convention or its protocols.

Directly precautionary measures may not remain precautionary over time. Advances in scientific capability, for instance, can resolve designed to address; in such a case, the measure is translated into a purely preventive strategy as the epistemological context of regulation is altered. Adjustments to moral context can have a similar effect. The commercial whaling moratorium, for instance, may have been instituted as a precautionary measure, but changing social mores may reinstitutionalize it as a straightforward moral imperative.[162] If environmental ethics evolve in such a way that pollution comes to be regarded a morally bad per se, rather than because of the disutility it may engender, then ironically the scope for precaution is consequently limited.[163]

A particularly significant form of direct precaution is that urged in the Bamako Convention, namely "clean production methods, rather than the pursuit of a permissable emissions approach based on assimilative capacity assumptions."[164] The Convention further states that:

In this respect Parties shall promote clean production methods applicable to entire product life cycles including: raw material selection, extraction and processing; product conceptualization, design, manufacture and assemblage; materials transport during all phases; industrial and household usage . . . clean production shall not include "end-of-pipe" pollution controls such as filters and scrubbers . . .[165]

Calls for clean production are predicated on the scientific inadequacies of emission standards based on assimilative capacity assessments of the environment in question. Assimilative capacity calculations require a predictive assessment of causal pathways within ecosystems; they therefore run counter to the inherent uncertainties of predictive science in areas of causal uncertainty and complexity, discussed in Section 1.4.3 above. Early criticisms of assimilative capacity calculations came from within the scientific community and tended to argue for the incorporation of a substantial margin of error into the determination of acceptable emission standards; levels of acceptability should be set by technical feasibility criteria – the use of "best available technology" – and therefore lower, and more precautionary, than would otherwise have been the case.[166]

Clean production takes these kinds of arguments further by taking account of "all of the material flows through society."[167] It has been defined by an expert group convened by UNEP's Industry and Environment office as: "the conceptual and procedural approach to production that demands all phases of the life-cycle of a product should be addressed with the objective of prevention or minimization of short and long-term risks to humans and to the environment."[168]

Moreover, a legal commitment to clean production will contribute substantially to meeting the environmental imperative of integrated forms of pollution control. The Siena Forum on International Law of the Environment has noted that:

The "sector by sector" approach, adopted in concluding conventions, often dictated by the need to respond to specific incidents, involves the risk of losing sight of the need for an integrated approach to the prevention of pollution and the continuing deterioration of the environment.[169]

The precautionary principle, even instituted on a sectoral basis, is unlikely ever to lose sight of that need. This is because it requires clean forms of production to be prioritized over alternative methods of regulation that might allow for unintegrated and exploitative employment of environmental sectors not directly covered by the legislation in question. Latter forms of regulation are premised on the possibility of meaningful comparisons of risk assessment across sectors and, as Jackson has forcefully argued that premise is highly questionable.[170] Where comparisons are meaningfully possible is between regulation not based on clean production and regulation that is; the latter will demonstrably reduce the scope for uncertain environmental impacts, and the former will not. Hence the precautionary principle gives rise to clean production which will in turn give rise to integrated environmental strategies, if only by default.[171] Clean production is a necessarily integrative form of pollution control.

The Environmental Protection Act 1990 (EPA) introduced integrated pollution control (IPC) in the U.K.[172] The Act makes it an offence to carry on a prescribed process in Great Britain except pursuant to an authorization granted by the appropriate authority and in accordance with any conditions imposed.[173] A condition implied into every authorization is that the best available techniques not entailing excessive cost (BATNEEC) will be used, inter alia,

> for minimising the pollution which may be caused to the environment taken as a whole by the releases having regard to the best practicable environmental option available as respects the substances which may be released.[174]

The EC Commission is currently working on their version of an integrated pollution control system, much influenced by the U.K. Act, based expressly on the precautionary principle. The Commission has decided that BAT will be the cooperative standard – without the NEEC, which is considered difficult in principle and in practice to apply.

A licensing system, which is truly integrated, will be a most effective method of institutionalizing precaution.

5. Conclusion

In earlier drafts of this paper we included a model definition. Events in international law-making on environmental matters have made that exercise somewhat superfluous – another definition is perhaps the last thing decision-makers need now. There is now so much precautionary language in international political/legal discourse on environmental protection that we would prefer to conclude by celebrating the precautionary principles contribution to a vital debate. As is cited above in 1.1.4. Agenda 21 the combination of burden of proof shifting and considerations of risk provide the precautionary principle with a powerful inquiry – "can you, the polluter show, that your activity poses no unacceptable risk to people and the environment?".

Notes

1 The authors wish to thank Tony Dymond, Barrister, for his invaluable assistance in discussing contested concepts and in preparing the text for publication.

2 See Cameron, J., Abouchar, J., "The Procautionary Principle: A Fundamental Principle of Law

and Policy for the Protection of the Global Environment", Boston College International & Comparative Law Review, Vol. XIV, No. 1., Winter 1991.

[3] O'Riordan, T. (1992) "The Precaution Principle in environmental Management", Centre for Social Economic Research on the Global Environment, GEC 92-03. See also von Moltke, K., "The Vorsorgenprinzip in West German Policy," Royal Commission on the Environment, 12th Report, Appendix 3, 1988.

Legislative references of the Vorsorgenprinzip include Bundesimmissionsschutzgestz (Federal Emission Control Act); the Atomgesetz 9 (Nuclear Energy Act), art 7, note 2, no 3; and the *Gesetz über die Umweltvertraglichkeitsprufung* (Environmental Impact Assessment Act).

[4] For a useful summary review of references to the principle, see Hey, E., "The Precautionary Approach and the LDC", paper prepared for the Secretariat of the LDC, August 1991, LDC 14/4, 4 September 1991. See also 20 Environmental Law and Policy, No. 3; and Freestone, D., "The Precautionary Principle", Ch. 2 in Freestone & Churchill (eds), "International Law and Global Climate Change" (Graham & Trotman/Martinus Nijhoff, 1991).

[5] U.N. Secretary-General, Annual Report on the Law of the Sea, November, 1990, UN Doc. A/45/721, cited in Freestone, D., "International Law and Sea Level Rise, Ch. 7 of Freestone & Churchill, supra note 3.

[6] See for instance Handl, G., "Environmental Security and Global Change: the Challenge to International Law", in Handl (ed) 1 Yb. Int'l Env.L. (1990). See also Gundling, L., "The Status in International Law of the Precautionary Principle", International Journal of Estuarine and Coastal Law, Vol. V, Ch. 3, 1990, and Nollkaemper, A., "The Precautionary Principle in International Environmental Law: What's New Under the Sun?", Marine Pollution Bulletin, Volume 22, No. 3, March 1991.

[7] Second International Conference on the Protection of the North Sea, London, 24–25 November 1987, Ministerial Declaration, issued by the Department of the Environment of the United Kingdom, April 1988.

[8] Baltic Sea Declaration, adopted at Baltic Environment Conference held at Ronneby (Sweden) 2–3 September 1990, IMO Doc. MEPC 30/22/5, Annex. The text is given in 1 Yb. Int'l. Env.L. (1990), pp. 423–429.

[9] Title XVI, Article 130r, section 2 of the Treaty of Rome as amended by Title II of the Treaty on European Union signed in Maastricht on February 7, 1992. The Maastricht agreement has not yet been ratified by all the signatories. At the date of writing, Denmark has decided, by referendum, not to ratify.

[10] Directive 90/219, Council Directive of 23 April 1990 on the contained use of genetically modified micro-organisms, O.J. L117/1 (1990), and Directive 90/220, Council Directive of 23 April 1990 on the deliberate release into the environment of genetically modified organisms, O.J. L117/15 (1990).

[11] Op. Cit. Principle 15.

[12] Op. Cit. Article 3(3).

[13] Op. Cit. Ch. 22 Para. 22.5(c).

[14] Preamble, paragraph 6 of the Montreal Protocol on Substances that Deplete the Ozone Layer, as amended at the Second Meeting of the Parties to the Montreal Protocol, London 27–29 June 1990, Doc. UNEP/Oz.L.Pro2/3, Annex II, p. 25. Text in 1 Yb. Int'l. Env.L. (1990) pp. 591–657. See also Adjustments and Amendments to the Montreal Protocol on Substances that Deplete the Ozone Layer, 30 International Legal Materials 1991, pp. 537–554.

[15] Recent Consultative Meetings of the Contracting Parties to the Convention on the Prevention of Marine Pollution by Dumping of Wastes and Other Matter [hereinafter, the London Dumping Convention] provide one illustration of this. See for instance paragraphs 4.1–4.19 and Annex 2 of the Report of the Fourteenth Consultative Meeting, 25–29 November 1991, LDC 14/16 30 December 1991. Additionally, the Oslo and Paris Commissions are currently reviewing their conventions; in both cases, the precautionary principle is listed in the reviews' terms of reference. See Hey, E., "The precautionary approach: Implications of the revision of the Oslo and Paris Convention" Marine Policy 15, July 1991, pp. 244–254. See also the "Report on the Activities

of the Oslo Commission", submitted by the Oslo Commission Secretariat to the Fourteenth Consultative Meeting, LDC 14/Inf.21, 4 November 1991.

16 See in particular the Comments on the Secretariat Paper LDC 14/4, submitted by the United States to the Fourteenth Consultative Meeting of the London Dumping Convention, LDC 14/INF.23, 4 November 1991. This argues that "the Convention, in its present form, incorporates a precautionary approach to ocean dumping," and that "it is largely because the Convention employs such an approach that it is unclear how express adoption of a precautionary approach would change decision-making under the LDC."

17 Gundling, p. 25, supra note 5.

18 Op. cit. pp. 29–30.

19 Handl, p. 23, supra note 5.

20 For the various national permutations of the Second North Sea Conference Ministerial Declaration, see Cameron and Abouchar, pp. 4–8, supra note 1.

21 International Maritime Organization: International Convention on Oil Pollution Preparedness, Response and Cooperation, 1990, and Final Act of the Conference, done at London, November 30, 1990, 30 I.L.M. 733 (1991).

22 Op. cit. p. 735.

23 European Economic Community-France-Morocco-Portugal-Spain: Final Acto of the Conference for the Protection of Coasts and Waters of the North-East Atlantic Against Pollution due to Hydrocarbons or Other Harmful Substances, and Accord of Cooperation, done at Lisbon, October 17, 1990, 30 I.L.M. 1227 (1991).

24 Alternative, and progressively more stringent, versions of this basic formulation, might include:

(a) *The precautionary principle stipulates that where the environmental risks being run by regulatory inaction are in some way (a) uncertain, but (b) non-negligible, regulatory action is justified.*

(b) *The precautionary principle stipulates that where the environmental risks being run by regulatory inaction are in some way (a) uncertain, (b) non-negligible, regulatory action is required.*

0

25 Nollkaemper, p. 107, supra note 5. (Nollkaemper responds differently to this same question.)

26 Gundling, pp. 26–27, supra note 5.

27 Cameron and Abouchar, p. 3, supra note 1.

28 Report by the OECD Secretariat, "The Role of Uncertainty in Decision-Making in the Area of Environmental Protection", ENV/EC/ECO(91)12. This was made available to Parties to the London Dumping Convention as "Dealing with uncertainty", LDC 14/INF.9, 10 October 1991.

29 Code of Conduct on Accidental Pollution of Transboundary Inland Waters as adopted by the Economic Commission for Europe at its 45th session (1990) by decision c.45, Doc. E/ECE/1225; ECE/ENVWA/16, Section III.7.

30 Op. cit. Section III.6.

31 The consensual scientific view is contained in the First Assessment Report of the Intergovernmental Panel on Climate Change, and its February 1992 Supplement. As the latter stresses (p. 6):

Findings of scientific research since 1990 do not affect our fundamental understanding of the science of the greenhouse effect and either confirm or do not justify alteration of the major conclusions of the first IPCC Scientific Assessment . . .

32 The seriousness with which potentially-affected parties are reacting to this threat is evident in, inter alia, the Malé Declaration on Global Warming and Sea Level Rise (reprinted in Molitor, Michael, R., "International environmental law: primary materials", 1991), the Fourth South Pacific Regional Environment programme Ministerial Intergovernmental Level Meeting: Ministerial Declaration on Environment and Development, Noumea (New Caledonia), 9 July 1991 (UN Doc.A/CONF.151/PC/87, Annex), and in the Symposium on Climate Change and the Future of Small Island states and Low-lying Coastal Developing Countries, New York, 14 February 1992. A summary of the latter was submitted to the Fifth Session of the Intergovernmental Negotiating

Committee for a Framework Convention on Climate Change, as INC/FCCC/None No. 37, 27 February 1992.

Working Group I of the IPCC (Science) has estimated a 65 cm rise in sea-level by the end of the next century, given the assumptions of IPCC Scenario A (Business as Usual). Working Group II (Impacts) estimates the rise at between 30 and 50 cm by 2050, and about 1 m by 2100. See also, "Possible adverse effects of sea level rise on islands and coastal areas, particularly low-lying coastal areas", Report by the UN Secretary-General to the General Assembly at its 46th Session, Doc. A/46/156, 19 April 1991.

33 "Climate Change and the Future", Keynote Address by D. De St. Jorre, Minister for Planning and External Relations, Republic of Seychelles, to the Symposium on Climate Change and the Future of Small Island States. This is reprinted in the summary of that meeting, supra note 40.

34 On the implications of feedbacks, see Leggett, J., "The New Politics of Energy".

35 Cited in Hey, "The Precautionary Approach and the London Dumping Convention", p. 16, supra note 3.

36 World Charter for Nature, G.A. Res 37/7, GAOR, Thirty-Seventh Sess. Supp. No. 51 (A/37/51).

37 See UNEP, supra note 14.

38 Statement to the General Assembly of Global Legislators Organization for a Balanced Environment (GLOBE) by H.E. Mr. Renagi R. Lohia, OBE, Permanent Representative of Papua New Guinea to the United Nations, on behalf of the Alliance of Small Island States, Washington D.C., 3 February 1992.

39 See, inter alia, the Final Ministerial Declaration of the Second World Climate Conference, supra note 12.

40 For instance, the Montreal Protocol talks of "taking precautionary measure . . . taking into account technical and economic considerations . . ." (supra note 15).

41 Houston Economic Declaration of the Seven Major Industrialized Democracies, paragraph 62, 11 July 1990. In an address to the National Governors Association in Washington D.C. (26 February 1990), US Secretary of State James Baker explained that "no regrets" implied that:

> . . . while the United States continues to support scientific research into the greenhouse effect, [we] are prepared to take actions that are fully justified in their own right and which have the added advantage of coping with greenhouse gases. They're precisely the policies [we] will never have cause to regret.

He noted further that the doctrine of "no regrets" would provide support for "actions which make economic and environmental sense regardless of the outcome of scientific disputes over causes and effects."

42 Thus the Paris Commission recommendation 89/1. 22 June 1989, reads:

> [T]he contracting parties . . . [accept] the principle of safeguarding the marine ecosystem of the Paris Convention area by reducing at source polluting emissions of substances that are persistent, toxic and liable to bioaccumulate by the use of the best available technology and other appropriate measures.

43 One relevant standard here is that of "best available technology not entailing excessive cost." See pp. 20–21 of O'Riordan, supra note 2.

44 A greenpeace poll of world climate scientists conducted during December 1991 and January 1992 shows that 45% of those surveyed believed a runaway greenhouse effect is possible if action is not taken to cut greenhouse gas emissions. Some 13% of those surveyed thought such an effect probable. The 400 scientists polled included all those involved in the 1990 IPCC report, as well as others who have published on issues relevant to climate change in "Science" or "Nature" during 1991 (Greenpeace International, Press Release, United Nations, New York, 20 February 1992).

45 O'Riordan, T., "The Precaution Principle in Environmental Management", CSERGE paper, supra

note 2, pp. 25–26. See also p. 23 for a game theoretical illustration of the linkages between science, economics, politics, as well as precautionary and non-precautionary responses to environmental threats.

46 Indeed, one potential precautionary procedure would be some form of insurance mechanism to compensate affected parties for externally caused environmental degradation. Such a scheme has been advance, for instance, in the context of possible sea-level rise. For further details, see Section 3.2.3b.

47 O'Riordan, ibid, p. 1.

48 For a more general argument to the same effect, see Sands, P., "The Environment, Community and International Law", Harvard Journal of International Law, Vol. 30, No. 2, Spring 1989.

49 Op. cit.

50 Final Ministerial Declaration, The Hague, 8 March 1990, p. 4. The entire document is reproduced in 1 Yb. Int'l Env.L. (1990), pp. 658–91.

51 Supra, note 6.

52 Gray, J.S., "Statistics and the precautionary Principle", Marine Pollution Bulletin, Volume 21, No. 4, pp. 174–176, 1990. See also Johnston, P. & Simmonds, M., "Reply to Gray", Marine Pollution Bulletin, Volume 21, No. 8, p. 402, 1990; and Gray's subsequent rejoinder to this in Marine Pollution Bulletin, Volume 21, No. 12, pp. 559–600, 1990.

53 Kuhn, T.S., "The Structure of Scientific Revolutions" (University of Chicago Press, 1970) p. 86. See also pp. 1–11 and pp. 82–86.

54 Jackson, T. & Taylor, P.J., "The Precautionary Principle and the Prevention of Marine Pollution," 1991, pp. 1–2.

55 OECD, "The Role of Uncertainty in Decision-Making in the Area of Environmental Protection", p. 5, supra note 36.

56 Bergen Conference, "Joint Agenda for Action", May 11, 1990, pp. 2 and 13. This document resulted from a series of collaborative NGO/governmental workshops held as a part of the Bergen process. It claims that it "represents a common view of representatives of Governments, business, voluntary environmental organizations, youth, science and labour on proposals to promote sustainable development which merit discussion, consideration or decisions by the various partners in the Bergen process, either jointly or separately" (p. 1).

57 Jackson and Taylor, pp. 6–7, supra note 69.

58 Sperling, K.R., "Protection of the North Sea" Balance and Prospects", Marine Pollution Bulletin, Volume 17, No. 6, pp. 241–246, 1986. See also Sperling, K.R., "The Dangers of Risk Assessment Within the Framework of the Marine Dumping Conventions", Marine Pollution Bulletin, Volume 19, No. 1, pp. 9–10, 1988.

59 Skolnikoff, Eugene B., "The Policy Gridlock on Global Warming" Foreign Policy, Summer 1990, p. 83.

60 Greenpeace International, Critical Review of GESAMP Reports and Studies No. 45 on Global Strategies for Marine Environmental Protection (1991), LDC 14/INF.29, 22 November 1991, p. 2.

61 Dewey, John, "The Public and Its Problems: An Essay in Political Inquiry" (Chicago, 1927), cited in Parson and Clark.

62 Parson, Edward, A. & Clark, William, C., "Learning to Manage Global Environmental Change: A Review of Relevant Theory", CSIA Discussion Paper 91–13, Kennedy School of Government, Harvard University, December 1991, p. 5. Note especially their sections 4.4 (Learning in Science) and 4.5 (Learning in Policy-Making), pp. 19–23. This paper provides an extremely useful review of learning theory, together with a comprehensive bibliography.

63 See infra 3.2.3d & 3.2.4.

64 "The prior justification procedure is a most rigorous application of the precautionary principle in that it places the burden of proof on the application to demonstrate that no harm will be caused to the marine environment." Freestone, p. 25, supra note 3.

65 Olson, supra note. See especially pp. 898–99. Olson also provides a detailed accounting of where the burden of proof has begun to shift within domestic US environmental law.

66 See note 9.

67 Treaty of Rome, Article 130r, section 2, as amended by the Single European Act, O.J., 1987, L.169, pp. 1–5.

68 Supra, note 9.

69 See Hey and Oslo Convention Secretariat Report, supra note 16.

70 See especially Articles 2–4. The entire CRAMRA text is reproduced in 27 I.L.M. 859 (1988).

71 See, for instance, Article 37 of the Fourth ACP-EEC Lome Convention, 15 December 1989, reprinted in 29 I.L.M. 809 (1990).

72 See the North Sea Continental Shelf Cases (West Germany v. Netherlands; West Germany v. Denmark), 1969 I.C.J. 3, 43–44.

73 See Hey, supra note 3; and Cameron & Werksman, supra note 80.

74 Supra note 17.

75 See Bin Cheng, "United Nations Resolutions on Outer Space: Instant International Customary Law", 5 Indian Journal of International Law 23 (1965). See also Cheng, "Custom: The Future of General State Practice in a Divided World", in Johnston and MacDonald, "The Structure and Process of International Law" 513 (1983); Kunz, "The Nature of Customary International Law", 47 American Journal of International Law 662, 666 (1953); Tunkin, "Remarks on the Juridical Nature of Customary Norms of International Law", 49 California Law Review 419 (1961).

76 Christol, C., "The International Law of Outer Space" 134 (1962). The International Geophysical Year lasted from July 1, 1957, through December 31, 1958.

77 Kirgis, p. 318, supra note 44.

78 Article 38.1 of the Statute of the International Court of Justice lists as a final source of international law, "judicial decisions and the teachings of the most highly qualified publicists of the various nations, as subsidiary means for the determination of the rules of law."

79 Freestone, p. 37, supra note 3.

80 Op. cit.

81 Louise de La Fayette, review of Freestone & Ijlstra (eds), "The North Sea: Perspectives on Regional Environmental Cooperation" (Graham & Trotman, 1990), in Netherlands International Law Review, Vol. 38, pp. 73–79, 1991.

82 McDougal, Myres, S. & Reisman, W. Michael, "The Prescribing Function in World Constitutive Process: How International Law Is Made", Yale Studies in World Public Order, Volume 6, Number 2, Spring 1980, pp. 249–284. See also Reisman, "International Lawmaking: A Process of Communication", Harold, D. Lasswell Memorial Lecture, Washington D.C., 24 April 1981, in "Order, Freedom, Justice, power: The Challenges for International Law", Proceedings of the 75th Anniversary Convocation of the American Society of International Law, 1981.

83 Reisman, ibid, p. 105.

84 Reisman, ibid, p. 108; see also McDougal & Reisman, ibid, esp. pp. 250–54.

85 Reisman, ibid, pp. 106 & 110.

86 Reisman, ibid, pp. 110–113.

87 Compare Reisman's informative example of the prescriptive emergence of the various unclear testing regimes; Reisman, ibid, pp. 114–119.

88 At Bergen, see Section 3.1.1a below.

89 This is how we would characterize the "no regrets" doctrine. It clearly seeks to capture the political force of precautionary language, since it is portrayable as a more activist response to prevailing uncertainties than a pure research strategy. At the same time, as Section 1.5.2 demonstrates, the conceptual and procedural differences are evident enough.

90 U.S. submission to the Fourteenth Consultative Meeting of the London Dumping Convention, supra note 77.

91 Op. Cit. Articles 7–14.

92 Resolution 687 (1991), Adopted by the Security Council at its 2981st meeting, on 3 April 1991, UN Doc. S/RES/687 (1991), 8 April 1991, Sections E.16-E.19.92.

93 See especially the far-reaching Nordic Convention on the Protection of the Environment, Stockholm, 19 February 1974, 13 ILM 591 (1974). See especially Article 3.

94 See Collins, L. "European Community Law in the United Kingdom" (1984), pp. 34–52. For cases,

see, inter alia, Case 152/84, Marshall v. Southampton and South-West Hampshire Area Health Authority, [1986] E.C.R. 723, 1 C.M.L.R. 688; Case 222/84, Johnson v. Chief Constable of the Royal Ulster Constabulary, [1986] E.C.R. 1651, 3 C.M. L.R. 240; Case C/188/89, Foster v. British Gas, [1990] E.C.R._, 2 C.M.L.R 833.

Furthermore, the European Court has extended the concept of direct effect to cover certain agreements between the Community and third party states. See Hauptzollampt Mainz v. CA Kupferberg & CIE KG, Case 104/81, [1982] E.C.R. 2641. See also Collins, ibid, p. 51.

95 Fuhr, Ormond, Roller & Gebers, Agenda for the Workshop "Access to Justice", Brussels, March 6, 1992, pp. 3–6. DG XI commissioned a report to advise in the drafting of the Access to Justice in Environmental Matters directive from the above authors and James Cameron.

96 EEA was established together with the European Environment Information and Observation Network by Regulation 1210/90, Council Regulation of May 1990, O.J. 1210.1 (1990).

97 Supra note 119.

98 See Chayes & Skolnikoff, "A Prompt Start: Implementing the Framework Convention on Climate Change", Report from the Bellagio Conference on Institutional Aspects of International Cooperation on Climate Change, 28–30 January 1992, circulated at the Fifth Session of the INC. This advocates an institutional strategy that "reflect[s] the overriding importance of transparency in international environmental regimes", p. 9.

99 1992 U.N. Framework Convention on Climate Change Article 12, see also Article 4(1) (a)&(b).

100 See, *inter alia*, Krasner, E. (ed), "International Regimes" (Ithaca: Cornell University Press, 1983); Haggard & Simmons, "Theories of International Regimes", International Organization 41 (1987), pp. 491–517; Keohane, R.O., "After Hegemony: Cooperation and Discord in the World Political Economy" (Princeton University Press, 1984).

101 For this view see Strange, S., "Cave! Hic Dragones: A Critique of Regime Analysis", pp. 337–54 in Krasner, supra note 116. For the classical realist and neo-realist foundations of this sceptical view see, respectively, Morgenthau, H.J., "Politics Among Nations: The Struggle For Power and Peace (5th edition, New York; Knopf, 1978), and Waltz, K., "Theory of International Politics" (1979).

102 This extreme view is most closely associated with Waltz, supra note 117. For a comprehensive examination of this position, see Keohane, R.O., "Neorealism and its Critics" (New York: Columbia University Press, 1986).

103 See, for instance, Gilpin, R., "War and Change in World Politics" (Cambridge University Press, 1981), p. 35.

104 Morgenthau, ibid, p. 286; Strange, ibid, p. 350.

105 Morgenthau, op. cit.

106 Noordwijk Declaration on Atmospheric Pollution and Climate Change (adopted at the Ministerial Conference on Atmospheric Pollution and Climactic Change, 5–7 November, 19 E.P.L. 229 (1989).

107 Global Environment Change Report, Vol. II, No. 10, 25 May 1990, p. 1.

108 Following section is based on Global Environmental Change Report, ibid, pp. 1–3. See also The Independent 14 May 1990 ("Doubt can justify action on pollution"), and ECO, 10 May 1990, Number 2, p. 1 ("US memo reveals fierce opposition to the precautionary principle").

109 ECO, op. cit.

110 ECO, op. cit.

111 See, *inter alia*, Olson, M., "The Logic of Collective Action: Public Goods and the Theory of Groups", Harvard Economic Studies, Vol. 124 (1971); Schelling, T., "Strategy of Conflict" (Harvard University Press, 1960); Axelrod, R., "The Evolution of Cooperation" (New York, Basic Books, 1984); Oye, D. (ed), "Cooperation Under Anarchy" (Princeton University Press, 1986).

112 Statements recognizing interdependence, both in academic and international political contexts, abound. Thus, Maurice Strong, UN Doc. A/C.2/SR.1466 (1972):

> The essential theme of the modern age is interdependence, the interdependence of all the elements which sustain . . . life on planet, of man with those elements, of the natural system with man's needs and aspirations and most of all man with man.

113 See Haggard and Simmons, supra note 123.

114 Skolnikoff, supra note 74.

115 NATO during the Cold War, for instance, constituted a cooperative and expensive endeavour designed to reduce the possibility of a particular event, namely external aggression. The chances of such an event coming about, even in the absence of worst-case preventative measures, were regarded as uncertain, but the costs, were it to do so, were perceived as unacceptably high. The collective action problem was thus resolvable; despite recurrent and acrimonious debates about burden-sharing, cooperation within the membership of the alliance was on the whole maintained. Uncertainties over possible defections from the regime, as well as over the effectiveness of particular cooperative security measures, were perceived to be negligible relative to the high-cost alternatives that non-cooperation might threaten. Free-riding was confined to the margins of an essentially cooperative arrangement.

116 See especially Axelrod, supra note 135. See also Axelrod and Keohane, "Achieving Cooperation Under Anarchy: Strategies and Institutions" in Oye (ed), supra note 135.

117 See Haas, P., "Saving the Mediterranean: The Politics of International Environmental Cooperation" (Columbia University Press, 1990).

118 Allott, P., "International Law and International Revolution" Reconceiving the World" p. 7. Allott asks: "Whodunnit? It was Emmerich de Vattel in his study with an idea."

119 Allott, "Eunomia: New Order for a New World" (Oxford University Press, 1990), section 16.19, p. 304:

When law becomes not merely the multi-lateralization of individual purpose but the universalization of social purpose, it transcends self-interest and becomes the self-creating of a society.

This, he argues, is a genuine international society, contrasting with what he terms the "phantom international society" which has been generated by a "partial social process . . . conceived as the interacting of the governmental public realms of the state-societies as units of willing acting, mathematical points with location but without extension . . ." (16.8). Allott provides a diagrammatic illustration of the "new self-ordering of the new international society," (16.103) and of the new international law of this latter (16.104). These are reproduced as Figures 1 and 2.

120 Handl, supra note 5.

121 Thacher, supra note 32.

122 GESAMP (IMO/FAO/UNESCO/WMO/WHO/IAEA/UN/UNEP) 1991: Global strategies for marine environmental protection. Rep. Stud. GESAMP (45).

123 Supra note 77.

124 See Jackson, T., "Waste Reduction and the Prevention of Marine Pollution" Alternatives to the Disposal of Wastes at Sea", submitted to the Ad Hoc Group of Experts on the Annexes to the London Dumping Convention, 15 January 1990.

125 Gundling, supra note 5.

126 See, for instance, the various recent amendments to the Barcelona Convention.

127 Inter alia, the Michigan Environmental Protection Act, Mich. Comp. Law, 691.1203 (1) (1989); the federal Resource Conservation and Recovery Act, 42 U.S.C. 6972(a)(1) (1982); and section 107 of the federal Comprehensive Environmental Response, Compensation, and Liability Act, 42 U.S.C. 9607 (1982).

See James M. Olson, "Shifting the Burden of Proof: How the Common Law Can Safeguard Nature and Promote an Earth Ethic", Environment Law, Vol. 20, pp. 981–915. See especially pp. 908–909.

128 Proposal for a Council Directive on civil liability for damage caused by waste, COM (89) 282 final, O.J. C251/3 (1989).

129 The 1979 UN/ECE Convention on Long-Range Transboundary Air Pollution provides a useful point of contrast, since it expressly states that it "does not contain a rule on State Liability as to damage."

130 Thus legal requirements for parties carrying out environmentally-hazardous activities to take out commercially-available private insurance, are not strictly precautionary. It is sometimes argued that,

once insured, such parties may lose an incentive to behave in a precautionary fashion, since they will no longer have to meet the financial costs of liability. Thus the 1969 Brussels Convention on Civil Liability for Oil Pollution Damage (text in 9 I.L.M. 45, 1970), which imposes upon bulk carriers of persistent oils the obligation to effect insurance against the liabilities imposed by the Convention, is not precautionary.

131 See Mans Jacobson, "The International Conventions on Liability and Compensation for Oil Pollution Damage", in "Oil Pollution Claims and Liability", Legal Studies and Services Limited, London, November 1989.

132 See Sands, P., "Chernobyl: Law and Communication" (Cambridge: Grotius, 1988). Paris Convention is laid out at p. 53, and the Brussels supplementary at p. 68.

133 The term is Handl's. See 1 Yb. Int'l Env.L. (1990), p. 19.

134 Supra note 119.

135 Council Directive 90/313 EEC of 7 June 1990 on the freedom of access to information on the environment, O.J. 1990 L 158/56.

136 Sands & Bulatao, "International Procedural Aspects of Atmospheric Protection: Environmental Impact Assessment and Access to Information", CIEL Background Papers on International Environmental Law, No. 2/1991, January 1991, p. 20.

137 In June 1990, Environment Ministers from central, eastern and western European governments met in Dublin and agreed to compile a register of pollution problems throughout Europe through the new Agency. See The Observer, 16 June 1990, p. 3; The Independent, 17 June 1990, p. 10.

138 Cited in Sands, supra note 62.

139 Section B II, paragraph 2, Supra note 155.

140 UNEP Governing Council Dec. 14/25, Goals and Principles of Environmental Impact Assessment, UN Doc. UNEP/GC/.14/17 (1987).

141 The New Assessment Procedure has been drawn up by the LDC's Ad Hoc Group of Experts on the Annexes to the LDC. The work was originally undertaken in response to a request by the Parties at the LDC's Tenth Consultative Meeting to "prepare any recommendations for alternative procedures for the classification and assessment of wastes to be dumped at sea which would afford better protection of the total environment against the adverse impacts of waste." For an explanation of the NAP's origins and drafting, as well as the complete text itself, see LDC.2/Circ.266, 19 June 1990.

For the Friends of the Earth International critique of the above, see LDC 14/INF.18.

142 See Oslo Commission Procedures and Decisions Manual, pp. C 3/86-E, 1/10/-8/10; pp. C A/86-E, 1/2-2/2; and pp. C 8/89, 1/5-5/5.

143 Hey, pp. 250–251, supra note 16.

144 Presumably unwittingly, given the long-standing Dutch commitment to the precautionary principle.

145 See supra I.4.3a.

146 1986 OECD Decision of 1986 on exports from the OECD of hazardous waste, C(86)64(final). Requires, *inter alia*, that exporters of hazardous waste demonstrate that the waste disposal facilities they intend to use are environmentally safe.

147 Paragraph 11 of the Final Declaration of the Third International Conference on the Protection of the North Sea.

148 Agreed at the Seventh Ordinary Meeting of the Contracting Parties to the Barcelona Convention; this particular stipulation related to products containing organophosphorus compounds, and was explicitly linked to the precautionary principle. See LDC 14/INF.15.

149 OSCOM Decision 89/1 (14 June 1989). Stipulates that a potential discharger of industrial waste not dispose of that waste unless they demonstrate that it is composed of "inert materials of natural origin" that would "cause no harm to the environment." In addition, discharges are permitted only where there are no "practical alternatives on land."

150 Both in Article 4 of the 1988 Convention on the Regulation of Antarctic Mineral Resource Activities.

151 Article 12 of the Convention for the Protection and Development of the Marine Environment of the Wider Caribbean Region, 24 March 1983 ("The Cartagena Agreement").

152 PREEPCOMM for International Seabed Authority and for the International Tribunal for the Law of the Sea, supra note. See especially Part VIII, Protection and preservation of the marine environment from unacceptable changes resulting from activities at sea, Articles 104, 105, 108.

153 Article 11(c) of the United Nations General Assembly Resolution 7 (XXXVII) World Charter for Nature (1982).

154 Fourth APC-EEC Convention signed in Lome, 15 December 1989, Article 37. Reprinted in 29 I.L.M. 809.

155 Supra, note 178.

156 Convention on the Prevention of Marine Pollution by Dumping of Wastes and Other Matter, 1972, 11 ILM 1294 (1972).

157 Op. Cit. p. 2. Additionally, The Preamble to the NAP p. 1 of the Annex to LDC.2/Circ.266 states that:

> Uncertainties in relation to assessments of impacts on the marine environment will need to be considered when applying the NAP and a precautionary approach must be taken to address these uncertainties.

158 Annex to LDC.2/Circ 266, p.2.

159 Op. cit., pp. 6–7.

160 Op. cit., p. 7, para. 5.2.9.

161 Op. cit., para 5.2.10.

162 See D'Amato & Chopra, "Whales: their emerging right to life", 85 American Journal of International Law 21 (1991).

163 See, for example, the "ecomorality" advanced in Kwisnek, E., "Earth or Consequences? Mythologizing the Earth Entity as a Way to Environmental Awareness", Duguesne Law Review, 1991.

164 Supra note 8.

165 Op. cit., Article 4(3)(g).

166 See Dethfelsen, supra note. See also Sperling, K.R., "Protection of the North Sea: Balance and Prospects" Marine Pollution Bulletin, Vol. 17, No. 6, pp. 241–246, 1986. There are elements in both these pieces that hint at endorsing some version of clean production, although they neither of them employ exactly this terminology.

167 Jackson and Taylor, p. 13, supra note 69.

168 Cited in Baas, Hofman, Huisingh, Huisingh, Koppert & Neumann, "Protection of the North Sea: Time for Clean Production," 1990, pp. 18–19.

169 Section 4; entire text reprinted in 1 Yb. Int'l Env.L. (1990), pp. 704–711.

170 Jackson, supra note 211.

171 Thus the International Association of Ports and Harbors are wrong to argue that the precautionary principle could be used to justify the dumping of dredged material at sea as the "environmentally preferable disposal method," even if in precautionary terms, marine disposal was preferable to other disposal strategies. Rather, the principle would argue for an elimination of the need for disposal by means of clean production, and hence obviate the need for these kinds of comparisons. See IAPH, "The precautionary approach and the management of dredged materials" LDC 14/INF.20.

172 Op. Cit. Part 1.

173 Op. Cit. Section 6(1).

174 Op. Cit. Section 7(7).

7. Policy Instruments, Public Deliberation and Evaluation Processes

JOAN SUBIRATS
Universitat Autónoma de Barcelona, Barcelona, Spain

In the beginning God created the heaven and the earth.

And God saw everything that He made. "Behold," God said, "it is very good".

And the evening and the morning were the sixth day.

And on the seventh day God rested from all his work. His archangel came then unto Him asking, "God, how do you know that what you have created is 'very good'? What are your criteria? On what data have you based your judgement? Aren't you a little too close to the situation to make a fair and unbiased evaluation?" God thought about these questions all that day and His rest was greatly disturbed. On the eighth day God said: "Lucifer, go to hell".

Thus was evaluation born in a blaze of glory. (Patton 1986, 1.)

1. Introduction

Evaluation can not only serve as a means of measuring or appraising the performance of policy instruments but also to foster consensus building among the different actors involved. This is done through a social learning process that involves discussing the criteria used to evaluate how effective the measures adopted are in terms of the problem or problems a specific policy is intended to address. In this latter sense, evaluation is also a policy instrument, particularly in the case of policies that require a change in human behaviour or where there is little consensus about which variables should be used to measure the effectiveness of the policy adopted.

In this paper I will not attempt to discuss the characteristic features of evaluation studies so much as the evaluation process itself, the way the various actors contribute to the evaluation, how this process relates to social learning systems and how the outcome of the evaluation can subsequently be used to reformulate policies and the way they are implemented. I will discuss mainly

Bruno Dente (ed), Environmental Policy in Search of New Instruments, 143–157.

environmental policies as they deal with a field in which these ideas are particularly applicable.

My working hypothesis is that institutional and cultural factors have a major influence on which policy instruments are selected, how they perform. It strikes me as unreasonable to maintain that the choice of any given policy instrument and the evaluation of its effectiveness should be based solely on technical considerations. The choice between economic or administrative measures in order to control pollution will be based on ideological beliefs and philosophical tradition as much as on strictly technical reasons. To put this very simply: the opinions of people who are in favour of regulating the market in order to control resources like air and water, heretofore scarcely marketed commodities, are people with a favourable opinion of regulatory devices while people who are reluctant to let market considerations influence any aspect of society will oppose such regulations.

Neither does it seem possible to maintain a clear distinction between objectives and instruments when the constant interaction between the two is obvious. It has been said that "demonstrating that there is a problem which can be attacked by one's favourite instrument is a very real preoccupation of participants in the policy process" (Majone, 1989, p. 117). Once the truth of this is acknowledged, it can safely be said that the performance of the instruments selected (and therefore their evaluation) will depend not only on their theoretical technical effectiveness, but also on the political, social and governmental environments in which they operate.

In this context evaluation is a weapon that all participants can use to reinforce their favourite points of view, but this apparent weakness (partiality) can become a strength or a way of building consensus, of charting progress, if evaluation is made from what might be termed a "multiple" or "pluralist" prospective.

2. On Evaluation

Although there is some confusion surrounding the term, "evaluation" can be said to be traditionally linked to analyzing the results of specific policies or programs and an essential part of the policy process. It has been said that "evaluation research" makes it possible to relate objectives to results in a perspective not unlike that of productivity analysis (Dente, 1989, pp. 75–76; Morehouse, 1972). Looking at it from this angle, evaluation should involve:

1) enumerating the programme's general aims and specific objectives;
2) discovering measurable indicators of these objectives;
3) collecting data about indicators for those people or things affected by the program and for an equivalent non affected control group;
4) and analyzing the data on what has been done and the controls introduced in the light of the program's initial aims and objectives (Weiss, 1972).

Taking these as their point of departure, a number of writers began attempting to give the evaluation process a scientific basis by introducing experimental or quasi-experimental techniques (Campbell-Stanley, 1966; Langbein, L.I., 1980; Rossi-Freeman, 1985). Critics of this approach have pointed out not only the difficulties involved in fully developing its initial hypotheses but also how little it reveals once it has demonstrated the difference between what was intended and what was actually done. These exercises in evaluation are somehow limited to viewing policy on its own terms. The value of what has been done is judged on standards inherent in the policy itself, which implies accepting the policy goals and ignoring other, external values which might question the "rightness" of this policy (Jones, 1992, p. 242).

(a) *The Limitations to the Classical Approach*

It is a well-known fact that the classic approach to evaluation requires a series of conditions which frequently render it impossible or make its final conclusions scarcely relevant. It assumes that the decision maker is an individual and acts as such. It assumes that his desires are consistent and that his preferences can therefore be assigned an order and the likelihood of certain things occurring evaluated. It assumes that the main thing is the outcome of the decision, not the process leading up to it. Intermediary or consensus-building factors are not part of the hardcore analysis. "Decisionists look upon policy problems as if they were puzzles for which, given clear goals and sufficient information, correct solutions always exist and can be found by calculation rather than by the exercise of political skills" (Majone, 1989, pp. 19–20).

The classic approach is somehow based on the assumption that it is possible to give scientific (objective) responses to the questions that comprise the evaluation process. But the very experience of evaluations with the greatest scientific pretensions reveal how the experimental ingredient in evaluation programs was gradually reduced or watered down until it reached the point where it became evident that "social reality differs substantially from physical reality" (House, 1982; along the same lines: Monnier, 1987). The institutional framework, the set of actors, the interests at stake varied according to the scenario, but intermediation and political commitment consistently outweighed pure program design. This indicates that people who stress the usefulness of evaluation, its ability to change behaviour and processes or who are seeking answers about values rather than just information about whether a given policy has met it objectives or not, would do well to start looking for a more plural (more "social") alternative to the necessary evaluation process.

(b) *Policy and Evaluation*

Public policy is inevitably the subject of political dispute and not just in terms of values ("Is this fiscal policy really fair?") but also in terms of the instruments used to implement these policies ("Are these really the most appropriate economic measures for increasing industrial production?") (Jones, p. 25). It is difficult to keep evaluations strictly separate from and uncontaminated by politics. There are a number of reasons for this: evaluations are inevitably part of the atmosphere of political decision-making that surrounds the program being evaluated; moreover the very content of the evaluation, its final value judgement, has a clearly political component; and lastly, evaluation can at worst be no more than a political instrument to support or criticize a particular program (Palumbo, 1987, p. 12). The problem is not so much how to reduce the invariably political content that is present to a greater or lesser extent in all evaluations, but how to avoid reducing the evaluation process to a mere ritual designed to ratify or attack a particular policy.

The definition of politics as "something that cigar-smoking males do behind closed doors" (Palumbo, 1987, p. 17) is unacceptable just as it is unacceptable to confuse politics with inter-party competition. It is in the interaction between actors and their resources (power, influence, authority) that the basis of the intermediation inherent to all politics is to be found. It is precisely the recognition of this plural basis that defines our approach to the evaluation process in which the inclusion of these multiple decision-makers and interests is a determining factor.

The intimate relationship between political scenario and evaluation process is illustrated by the fact that every program or action to be evaluated has its political decision-makers; its sponsors; the administrative personnel who have been involved in it; staff, clientele and interest groups who have in some way been affected by it. Evaluation is produced by and related to politics and becomes part of the political decision-making process. But, above all, evaluation is somehow "political" in that it makes a certain value judgement which, though attempting to be objective, nevertheless involves taking a particular stance (Weiss, 1987). One of the points I will examine in this paper is how to deep from turning evaluation into a mere "advocacy game" without necessarily accepting the legendary attributes of the scientific approach.

(c) *Pluralist Evaluation*

A different concept of evaluation has been taking shape for some time now and it is one that does not attempt to evade the political aspect inevitable in evaluation or neglect the need to acknowledge the social interaction inherent in all public action programs. Instead it attempts to combine the presence of the principal actors in the evaluation discussion with the necessary subsequent use of their findings. Although this kind of evaluation has not as yet been given a definite label, "pluralist evaluation" would appear to describe what other authors refer to as "naturalistic evaluation", "stakeholder evaluation", "utilization-focused evaluation" or "multiple evaluation" (Guba-Lincoln, 1981; Bryk, 1983; Patton, 1986; Majone, 1989).

All these approaches share a certain scepticism about the possible usefulness of evaluations that claim to be partial and objective. the pluralist concept maintains that it is not enough to base evaluation on purely empirical information without taking personal values and opinions into consideration. "In fact, values and opinions count a great deal in evaluation, not only because of the ambiguity of the outcomes in practice . . . but even more because of inescapable disagreements about the kind of evaluative criteria that are meaningful, fair or politically acceptable in a given situation" (Majone, 1989, p. 168). These ambiguities or problems in focus cannot be resolved simply by employing more and better measurement techniques, but also require a great deal of public discussion and persuasion. Analytical instruments, with their varying degrees of objectivity, would make more sense and become more operative were evaluation standards first defined in this public deliberation process.

The truth is that the actors involved in any public action program are always quick to offer their critical opinions. In this sense any evaluation made by specialists is no more than just another opinion. Policies and policy instruments used in public action programs are constantly appraised and evaluated in the light of the particular logic and outlook of the people directly or indirectly involved. And these multiple approaches are the very core of the policy design process in a pluralist society.

3. Evaluation as A Policy Instrument. Evaluation as A Social Learning Process

This paper will attempt to defend an approach to evaluation which does not simply emphasize the extent to which the internal objectives of the policy itself are or are not met, but acknowledges the discussion about the basic values of the policy itself, the analytical process and the findings obtained and aims to involve the principal actors in the evaluation process.

Evaluation is thus understood to be one more example of social pluralism and an admission that a number of different value systems exit. Viewed from this angle, evaluation becomes part of the process whereby the members of society learn how to solve collective problems (Stake, 1975; Monnier, 1987). Examining the program to be evaluated becomes less a matter of discovering the "truth" about the effects of a particular program and more a way of revealing its different "truths" in an attempt to strike a balance which will be greater than the simple sum of the opinions of the individuals involved. As has recently been stated, "possibly the most effective single way to reduce impairment is to get into circulation a greater variety of messages, (a) competition of ideas" (Lindblom, 1990, p. 293).

Emphasizing evaluation as an instrument for shaping confidence and creating consensus means involving everyone affected by the program or policy to be evaluated in an attempt to gradually reach the aimed-for "objectivity" by getting both specialists and the members of the general public to jointly discuss and explore the ramifications of the program or policy rather than simply providing a summary of some scientific premises that are always questionable. This would improve the policy being advocated, submitting it to a specific form of "multiple advocacy" (Majone, 1989, p. 40).

The evaluator would not act alone, arbitrarily deciding whether the program under discussion is good or bad. Instead, he would act as a mediator between the different opinions. His scientific and technical background would not make him an objective and impartial judge, but would enable him to act as a negotiator, evaluating the need for information, answering questions, explaining consequences and attempting to reach an agreement about criteria and priorities (Ballart, 1992).

(a) *The Limitations of An Exclusively Technical or Professional Approach and the Social Learning Process*

It is not my intent to question the importance of professionals in any field nor do I question the important role that evaluators can and do play in any process aimed at determining the usefulness of a particular policy. It is obvious that our society is increasingly dependent on specialists in a variety of fields. But, as our dependence on specialists grows so does our scepticism about their true capabilities and knowledge. We are sceptical because we have seen how the freedom of action of individual specialists or professional's is sometimes compromised. We are sceptical because we have seen certain theories or provisions backfire or because these same specialists have failed to foresee their effects. As if that were not enough, we are constantly hearing about partly or wholly contradictory theories and studies, all presented by the most highly qualified individuals or teams. As a result certain sectors of the public have joined consumer and other organizations in order to protect themselves from the corporatism of many professional associations. Moreover, there has been a considerable increase in the number of clients, patients, customers and other members of the public who take legal action against specialists.

This may seem incongruous in a society where specialization is becoming the general norm. Science and its technological applications have developed to a point that would have been unimaginable just a few years ago. Today the welfare and economic development of any country is commonly gauged by the extent of its "science and technology". And yet there are grave doubts about using all this scientific potential to resolve increasingly complex social problems. These doubts stem from the very errors, subordinations, weaknesses and inconsistencies of scientists and their solutions and are aggravated by a democratic system of collective decision-making that is difficult to reconcile with technocratic logic or ideas that bring to mind a scientifically guided society.

It has been observed that nowadays people are more aware of reality, they have more information and technological expertise and all this makes decision-making more complex. As reality becomes more complex and chaotic, professional skills become increasingly relative (Schön, 1983). This makes it necessary to know how to deal with indeterminate positions and conflicts of values. Complexity, instability, uncertainty and conflicts in value systems cannot be avoided or solved by applying rational techniques that attempt to provide specific solutions to well-defined, standard problems. Some practitioners manage to handle these complex situations by applying a vague mixture of intuition, skills and knowledge that results in a sort of "knowing-in-action" that is often hard to explain but which is based on the process of acting, accumulating the experience of these actions and developing a special sensitivity that enables them to detect the "winning habits" in a "reflection-in-action" process that can be synthesized as professional expertise (Schön, 1983, p. 21). These mixtures of intuitive judgement and knowledge of the subject in question have been described as "non-logical processes" (Barnard), "the art of judgement" (Vickers) and "tacit knowing" (Polanyi).

This sort of insight, hitherto related to basically individual actions, could perfectly well be applied to organizations, institutions and groups such as policy communities and policy networks, which though less structured are no less tangible. Here too it should be possible to learn from experience, from the interaction between the different members or factions of a particular group. If evaluation practices were given a pluralist focus and the various actors affected by the policy in question encouraged to express their opinions, these practices would truly become instruments of social learning. Were resources like knowledge and trust (Klok, 1992) to be shared, both belief and disbelief would be expressed, winning habits shaped and situations from which something could be learned created, thereby giving the evaluation process its fullest meaning as a policy instrument.

Evaluation could thus become a formal occasion for examining the way specific programs of action have been experienced by the people affected and discovering what they believe to be the successes and failures of these programs. In a situation where pluralism is accepted and there is a real desire to improve the operating environment, evaluation is a summing-up of the social learning resulting from the interrelations of the policy community. Policy errors will not be condemned, but will be viewed as sources of information and clues to subsequent actions to be taken.

> In this model, citizens, functionaries, social scientists and other experts do what they have learned and they learn what they have done. (Lindblom, 1990, p. 219)

(b) *Citizens as Analysts, Analysts as Citizens*

When a particular policy is classified as a success or failure, this frequently means that it has been viewed from a narrow managerial focus, more concerned with meeting internal policy goals of exercising effective administrative control than with the program's ability to respond to the needs of the various individuals and groups affected. It also means that valuable information for political decision-makers is being ignored. Moreover, efficiency and effectiveness, which are typical values of evaluation studies, often clash with the attitudes of policy implementers, who are more concerned with the professional skills and decision-making powers of the specialists who designed the policy.

Many public policies currently being implemented are based on the assumption that they can and will change the public's style of operations or their scale of preferences. In some cases public policies are expressly aimed at certain groups of citizens while others affect the public as a whole. In these cases, encouraging public participation not only expresses a general desire to make life more democratic, but is utterly essential if policies are to meet with greater success.

Deciding who should participate and how naturally poses problems. The people implicated in a particular policy (politicians, the experts or officials who implement the policy, special interest groups, individual citizens, outside experts . . .) are the people who should be involved in any pluralistic evaluation process. The process will then strike a balance between standard professional evaluations with their almost exclusive emphasis on expert opinions and the criticism that any action taken by the public powers provokes in one sector or another of society. All the individuals or groups affected have different opinions about the evaluation criteria to be used and the emphasis to be placed on particular aspects of the policy process. General standards of performance, such as legality, legitimacy, economy, effectiveness, efficiency and the ability to provide responses to social needs are the specific concern of certain of the actors involved (politicians, judges, consumers . . .), but it should also be acknowledged that each of these criteria can be more easily applied to a particular part of the policy process than to the process as a whole (for example, legality is more aptly linked to process, economy to inputs, etc.) (Majone, 1991).

But these multiple criteria mirror reality and must form part of the evaluation process, particularly if the aim is to subsequently apply the conclusions drawn, which are not so much the result of an agreement on what has been achieved, but of how much the policy community has learned about the problems revealed and the mistakes made.

The general criteria Wildavsky listed years ago (Wildavsky, 1979, p. 253) still apply. The policy community must understand what is in it for them, recognize the differences between small and large changes, and be involved continuously so that they can learn from experience (Wildavsky, 1979, p. 253). The ability to gradually shape a policy community that is sufficiently open yet specialized and that recognizes disappointment as a constant policy test is the best way to insure improved policy performance. And evaluation, as an opportunity to discuss objectives and results, may well be the most suitable instrument for doing this.

(c) *Using Evaluation*

All this is easier said than done, but if it can be achieved then evaluation will certainly be used to improve the program involved. Criticisms of the evaluation process are generally aimed at:

- the weakness of the methodology used in the evaluation. Often the procedure employed in evaluating a particular program causes the validity and credibility of the evaluation itself to be called into question.
- irrelevance. The outcome of an evaluation may be inadequate at a particular moment or may simply fail to provide the decisionists with any new information.
- infra-utilization. The information obtained is either not made public or not subsequently acted upon.

The idea of evaluation continues to be intrinsically linked to reveal whether or not a particular policy has had the hoped-for results. As mentioned earlier, evaluation tends to be based on the decisionists' own terms and implies that the decision-makers have decided how things should be done and subsequently taken actions designed to get results. Evaluation is intended to examine these results and indicate the extent to which the aimed-for results have been achieved. It is assumed that evaluation is an objective, internally coherent process when in actual fact the evaluators' very objectivity and supposed impartiality can be the reason underlying problems in getting results.

The general impression is that any short-term evaluation designed to classify a policy or program as a success or failure tends to conclude that the policy or program involved has had little effect. Because actions involving public powers are so complex, evaluation must be made less "peripheral" and viewed more as a learning tool than an objective measure of performance. For this reason it is absolutely essential that both the different levels of public powers involved in policymaking and the people who will be affected by these policies take part in the evaluation process.

There are so many different evaluation criteria and so many standards of accountability that the analyst cannot be expected to judge something that belongs more to the field of politics than to the realm of analysis. Instead, his role is to provide bases for discussion that are as appropriate as possible. "The need today is less to develop 'objective' measures of outcomes – the traditional aim of evaluation research – than to facilitate a wide-ranging dialogue among advocates of different criteria" (Majone, 1989, p. 183).

Rather than limiting us to a few pre-defined analytical categories we could take advantage of the different viewpoints of the people involved, their different sources of information and their different experiences with particular policies in order to discover effects and connections, links between processes and outcomes and examine how different factors interact with time.

Focusing more on what people really do, the real implications of their problems, their basic situations and the roles they play; charting the way they interact, their strategies and arguments; accumulating and compiling their experiences; studying their basic assumptions and comparing facts with these arguments and assumptions can all combine to form the basis for evaluation as a consensus-building instrument and provide a guarantee that it will be increasingly used as an instrument for improving policies (Hellstern, 1985).

4. Pluralist Evaluation, Public Deliberation and Environmental Policy

(a) *General Features*

It has been said (Bosso, 1987) that environmental policy is much more "resistant" to becoming part of an accepted routine than are other policies and that environmental policy is therefore invariably more "political". This may in part explain why in some countries there is so little tradition of environmental policy (viz. southern Europe) and why it is often considered a "mobilizing" policy (Olsen *et al.*, 1982, pp. 50–51). This means that there is a bargaining arena and a structure of the agenda that we cannot consider fixed, with a fairly high level of conflict, because of the disagreement over goals and casual theories as well as over procedural rules of the game. At the same time, the lack of consensus among the increasingly numerous experts in the field is gradually giving the whole field of environmental policy a reputation for fluidity, openness to change and renovation. There is a lot of conflict between the different groups but the internal cohesion of the most active of them is on the rise. It is difficult to talk about stable and structured policy communities in countries like Italy (Lewansky, 1990, p. 310) and Spain (Aguilar, 1992).

In addition, institutional responsibility is highly fragmented, with various levels of government and the public administration being simultaneously involved (either actively or passively).

In Europe alone, the extent of environmental awareness differs greatly from one country to another. The EC and other supra-national agencies design policies and enact regulations that have little relation to the priorities on the national, regional or local agendas of certain countries. numerous institutions split the responsibility for environmental issues and for developing specific regulations to control environmental deterioration.

These general features make environmental policy an appropriate field in which to experiment with pluralist approaches to evaluation that would make it possible to improve environmental action programs and control the application of specific regulations. As we all know, the actors most frequently involved in environmental policy (experts, managers, judges and jurists, economists, ecologists, politicians, the business community, the sectors of the community who are directly affected, the general public) often have radically different viewpoints.

Moreover, the general public, governments and experts are all increasingly concerned about the environment and this has made more information available, but is has also increased uncertainly about acceptable risk levels and shed doubt on forecasts about how certain variables are likely to develop (Hawkins-Thomas, 1989). The abundance of available data, the proliferation of studies and forecasts have only increased uncertainty and pointed up the fact that environment is not a linear system. Environmental policy has been labelled a policy of scarcity. This scarcity is sometimes artificially created by scientific or scientific and social consensus. The way society perceives scarcity is important. Equally important are information and public awareness of the gravity of the current situation and the need to enact urgent measures to stem environmental deterioration.

Although increased social awareness of environmental problems has led to increased controls, it has done little to alter the conduct of the people who cause pollution. If pollution is to be controlled, then industries must be responsible for controlling their own operations and introducing environmental protection measures into their production processes. Social pressure and public concern over the environment can be vitally important in changing industrial attitudes.

All this would seem to indicate that evaluation can be effective as an instrument for consensus-building and a means to improving environmental policies within a framework of public participation and social learning.

(b) *Evaluating Environmental Policies*

There is a great deal of doubt about the true effectiveness of recent regulations and how much environmental policies have actually done to change things.

All thoughts of evaluating environmental policy tend to be rooted in economic concerns or linked to civil engineering traditions. Cost-benefit analyses have demonstrated that political factors are crucial in determining costs and that the analysis technique was being used in a way that "looked neat, and seemed neutral to the naked eye. Reducing policy to numbers made people feel comfortable; the process gave a sense of clarity to issues that had heretofore defied easy definition" (Tolchin, 1987, p. 279). But in actual fact, the more qualitative elements that could pinpoint benefits and define impacts are played down because they are considered too complicated to measure or because they are indivisible.

Moreover, the "Environmental Impact Evaluations" recommended by EC Directive 85/337 as a preliminary to any public action that will in any way affect the territory, are highly technical and use standards of measurements that are often questionable and do not take the opinions of the people affected or the public in general into account even though reference is made to "the aptness of the social environment" for the decision being made (Gomez Orea, 1988). "Public participation" takes place at the end of the analytical process and is strictly a formality, similar to the "public participation" in city planning measures.

(c) *Evaluation as An Environmental Policy Instrument*

Environmental policy is closely linked to regulatory measures enacted by specialized agencies connected with the different actors involved. Policy making and discretion are highly interdependent, particularly inasmuch as it is so difficult to define so-called risk situations in an objective or generally acceptable way (see Bosso, 1987).

The interaction of policy makers, experts, bureaucrats, interest groups, users and the public should therefore be instrumental in measuring the extent of adherence to regulations and evaluating their effectiveness. The uncertainty surrounding risk situations and the limitations of science and technology recommend the use of public deliberation to determine what risks are acceptable and what kind of technological solutions, regulations or discretionary measures will be tolerated (Hawkins-Thomas, 1989, p. 273).

Processes based on the techniques of "utilization focused evaluation" (Patton, 1987, p. 122) or pluralist evaluation (Monnier, 1987) can provide an adequate framework for intermediation processes that help confer legitimacy on the measures adopted. Indeed, environmental policies are often based on the relationship between a physical or biological reality that is felt to be deteriorating or endangered and what is scientifically desirable. All this is part of a process whereby the different political agents make the corrections they consider necessary from their ideological viewpoints or on the grounds of social consensus. Evaluation is presented as a chance to discover whether things have improved or not, but the definition of "improvement" will differ depending on the viewpoint involved. Natural science may furnish new information but this information will be subject to different interpretations. The more unanimous the scientific facts, the greater the possibilities of reaching an agreement. Thus there is a direct, though not always straight, line between scientific facts and social convictions. Here too pluralist evaluation processes could help reinforce social convictions thereby increasing the chances that the measures adopted will be adhered to.

In terms of environmental policies, the relationship between scientific certainty and social certainty is conclusive in gauging how successful certain measures (designed to change social comportment, practices that cause pollution, etc.) will be. Myriad social interest, the disparity of criteria applied or positions adopted when judging whether the environmental policy measures enacted are correct or not, should increase the likelihood of social debate. And the existence of different schools of thought and different parameters will not infrequently keep scientific discussion alive (see the Asarco-EPA case; Scott, 1988). (See Table 1, where, position A is very infrequent in environmental problems, while position B allows a political decision that stops social debate, and position C offer some room to a formal evaluation because the social agreement on the non controversial nature of the issue; position D, probably the most frequent situation in environmental policies, fits very well in our pluralist approach of evaluation as instrument.) Open, pluralist evaluation processes will quite possibly prove to be a suitable way to confer legitimacy on specific environmental policies and will serve as an instrument for social learning in which scientific facts will be given their just due and improvements or deterioration in the behaviour of the public concerned will be clearly revealed, thus shaping a common ground for attempting to improve the definition and implementation of environmental policies.

Although evaluation of environmental policies as described here should take account of the different opinions of the members of society, it should not under-

Table 1. Scientific certainty and social certainty

Scientific certainty	Social certainty	
	yes	no
yes	A	B
no	C	D

estimate the facts presented by scientists and other experts. By not ignoring the way society receives these facts and alters its behaviour to fit the circumstances, the evaluation process should even be able to improve on these scientific facts, serve as a social and organizational learning tool and be instrumental in improving the environment.

References

Aguilar, S. (1992) *Políticas Medioambientales y Diseños Institucionales en España y Alemania: La Comunidad Europea Como Escenario de Negociación de una Nueva Area Política*, PhD Thesis, Madrid (mimeo).

Ballart, X. (1992) *Evaluación de Programas y Servicios Públicos*, INAP, Madrid.

Boss, C.J. (1987) *Pesticide and Politics*, University of Pittsburgh Press, Pittsburgh.

Browne, A. and Wildavsky, A. (1987) What should evaluation mean to implementation, in Palumbo (ed), *The Politics of Program Evaluation*, Sage, Newbury Park, pp. 146–172.

Bryk, A.S. (1983) *Stakeholder Based Evaluation*, Jossey Bass, San Francisco.

Campbell, D.T. and Stanley, J.C. (1966) *Experimental and Quasi-Experimental Designs for Research*, Skokie, Ill., Rand-MacNally.

Cronbach, C.J. (1980) *Toward Reform of Program Evaluation*, San Francisco CA, Jossey Bass.

Dente, B. (1989) *Politiche Pubbliche e Pubblica Amministrazione*, Rimini, Maggioli.

Gomez Orea, D. (1988) Evaluación de Impacto Ambiental, in *Ciudad y Territorio*, n. 75, pp. 5–32.

Guba, E.G. and Lincoln Y.S. (1981) *Effective Evaluation: Improving the Usefulness of Evaluation Results Trough Responsive and Naturalistic Approaches*, Jossey Bass, San Francisco.

Hawkins, K. and Thomas, J.M. (1989) Making policies in regulatory bureaucracies, in Hawkins-Thomas (eds), *Making Regulatory Policies*, University of Pittsburgh Press, Pittsburgh, pp. 3–30.

Hellstern, G.M. (1985) Assessing evaluation research, in Kaufmann-Majone-Ostrom (eds), *Guidance, Control and Evaluation in the Public Sector*, De Gruyter, Berlin, pp. 279–312.

House, E. (1982) Preface to *Evaluation Studies*, in Evaluation Review Annual **7**.

Jones, P. (1992) Evaluation, in Harrop, M. (ed), *Power and Policy in Liberal Democracies*, Cambridge University Press, Cambridge, pp. 241–262.

Kelman, S. (1981) *What Price Incentives?*, Auburn House, Boston.

Klok, P.J. *A Classification of Instruments for Environmental Policy*, ESF, mimeo.

Langbein, L.I. (1980) *Discovering Wether Programs Work: A Guide to Statistical Methods for Program Evaluation*, Glenview, Ill, Scott-Foresman.

Lewansky, R. (1990) La politica ambientale, in Dente (ed), *Le Politiche Pubbliche in Italia*, Il Mulino, Bolonia, pp. 281–314.

Lindblom, C.E. (1990) *Inquiry and Change*, Yale University Press, New Haven.

Majone, G.D. (1989) *Evidence, Argument and Persuasion*, Yale University Press, New Haven.

Majone, G.D. (1991) Public policy beyond the headlines, *EUI Working Papers in Political and Social Sciences*, European University Institute, Florence.

Monnier, E. (1987), *Evaluation de l'Action des Pouvoirs Publics*, Economica, Paris.

Morehouse, T.A. (1972) Program evaluation social research versus public policy, in *Public Administration Review*, pp. 868–874.

Palumbo, D.J. (1987) Politics and evaluation, in Palumbo (ed), *The Politics of Program Evaluation*, Newbury Park, Sage, pp. 12–46.

Patton, M.Q. (1986) *Utilization-focused Evaluation*, Sage, Beverly Hills.

Patton, M.Q. (1987) Evaluations political inherency. Practical implications for design and use, in Palumbo (ed), *The Politics of Program Evaluation*, Newbury Park, Sage, pp. 100–145.

Rossi, P.H. and Freeman, H.E. (1985) *Evaluation: A Systematic Approach*, Beverly Hills, CA, Sage.

Schön, D.A. (1983) *The Reflective Practitioner*, Basic Books, New York.

Scott, E. (1988) Managing environmental risk: The case of Asarco, *Case Program. Kennedy School of Government*, Harvard University, Cambridge.

Stake, R.E. (1975) *Evaluating the Arts in Education: A Responsive Approach*, Merrill, Columbus.

Tolchin, S.J. (1987) The political uses of evaluation research: cost-benefit analysis and the cotton dust standard, in Palumbo (ed), *The Politics of Program Evaluation*, Newbury Park, Sage, pp. 249–270.

Weiss, C. (1972), *Evaluation Research*, Englewood Cliffs, NJ, Prentice Hall.

Wildavsky, A. (1979) *The Art and Craft of Policy Analysis*, Macmillan, London.

8. Mediation as a Policy Instrument for Resolving Environmental Disputes

With special reference to Germany

HELMUT WEIDNER
Wissenschaftszentrum Berlin für Sozialforschung (WZB), Berlin, Germany

1. Introduction

In past years some areas of environmental protection have proved to be particularly resistent, or even allergic, to any attempts by environmental policy-makers to exert control. This includes large industrial and public projects in virtually all democratic industrialised countries where people are afraid of their negative consequences on health and the environment. Because as a rule they are facilities and infrastructural projects which are crucial to the functioning of an industrialised society (projects such as landfills, waste incineration plants, airports, motorways, dams, power stations) and which are considered to be fulfilling one of the state's prime responsibilities ("precautionary policies" to safeguard the economy), disturbances which may be ecological in origin also cause major political and societal conflicts in which powerful economic interests and state institutions are also actively involved. The projects are opposed not only by members of the public who could be negatively affected by them and environmental organisations but also more and more by institutions involved in local administration (local public administrators and politicians), which makes "hard" enforcement of state decisions using the conventional instruments considerably more difficult than if it were a case of dealing "merely" with societal representatives and their organisations. Opposition from parties which understand all the intricacies of the political and legal system must therefore also be taken into account.

If, as is increasingly the case, the project is not dropped due to these conflicts, its realisation often becomes a very time-consuming and expensive business – the original objectives are watered down, the conflicts are shifted to the implementation phase which in turn causes "enforcement deficits" (a decision which has been legally and politically sanctioned still requires implementation involving a whole series of individual documents) and the groups involved become embittered, not to mention the destructive effect on social relations within the town or region affected. In brief: traditional instruments at the disposal of governments to imple-

Bruno Dente (ed), Environmental Policy in Search of New Instruments, 159–195.

ment their public policy objectives and responsibilities are proving in major environmental disputes to be increasingly unsuitable for achieving solutions which are not only economically viable, but also compatible with the environment and the wishes of society.

It is quite obvious that the sharp increase in the economic, social and political costs of using traditional political instruments to deal with environmental conflicts is not simply a temporary phenomenon in the course of an issue-attention-cycle, not just the latest wave of protest ("environmental hysteria", "pollutant of the week") which manifests itself in a few satiated opulent societies but more a new secular trend in the most advanced industrialised countries (for instance,the so-called G-7 countries). None of the groups involved in the conflicts is happy with the situation. This is evident from the reciprocal accusations made. The accusations leveled at industry are: failure of the market, ignoring the environment as an economic principle; the environmental groups are accused of hysteria and inability to enter into dialogue with others; politicians and public administrators are criticised from all quarters as lacking impartiality and being incapable of taking decisions; critical scientists are even talking in terms of the *failure of the state* to protect the environment.

Everyone complains of the lack of appropriate instruments for mastering the challenge to implement effective environmental policies. This along with the pressing nature of the problem itself creates an openness towards unconventional instruments and encourages (by virtue of necessity) a willingness to take new approaches to dealing with environmental disputes. In any case a constantly growing increase in alternative forms of conflict management can be identified in certain countries, particularly within the governmental and industrial structures. Some countries even have many years of practical experience behind them: what is known as alternative dispute resolution is now well established in the U.S.A., Canada and Japan. In Europe, as I shall go on to point out, it leads only a shadowy existence; here we are still almost completely taken up by listening to the debate "unleashed" in the U.S.A. on economic environmental instruments and by considering putting them into practice, albeit hesitantly.[1]

2. Alternative Dispute Resolution Procedures

In the U.S.A. decision-making processes which complement traditional (formal) procedures are called "alternative dispute resolution procedures or approaches". The abbreviation ADR is commonly used. As the term "alternative" could give rise to misunderstandings, I should like to follow the example set in relevant literature on the topic and expressly point out that it does not refer to traditional procedures for conflict resolution or decision-making (such as parliamentary procedures or those undertaken by law courts or the public administration) being *replaced* but *complemented* by these new techniques. The alternative procedures can be used prior to, in parallel with, or subsequent to conventional procedures and other instruments or can be linked to them.[2]

All the various types of alternative procedure have one thing in common: they are intended to settle disputes – not necessarily conflicts – through negotiation. Participation in the procedures is voluntary. The hierarchical structure of conventional procedures has been almost entirely abandoned: representatives of public administration, for example, do not have any special authority to exert control or take decisions. The objective is to find a solution based on consensus, not a unilateral decision handed down from above that is binding whether the people concerned agree or not. These procedures are usually informal, i.e. the people taking part draw up the rules and regulations themselves. A certain tendency to make them subject to regulations can, however, be observed, in two senses: legislation of different kinds is being drawn up for alternative procedures and general standards are being put forward by professional institutions and associations (cf. The contributions of Sander and Gardner in Hoffmann-Riem and Schmidt-Aßmann, 1990).

A classification system prepared by Susskind and Madigan (1984, pp. 180 ff.) lists the following different types of procedure for settling disputes; the criteria for the classification is the "degree of activity" with which independent people "intervene" in the procedure.

Classification of alternative procedures for settling disputes

1. Unassisted negotiation
2. Facilitated policy dialogue
3. Collaborative problem-solving
4. Passive (or traditional) mediation
5. Active mediation or mediated negotiation
6. Non-binding arbitration
7. Binding arbitration
8. Adjudication

Source: Susskind and Madigan (1984)

In the meantime there have been numerous proposals for systematically classifying approaches to conflict management (cf. Susskind and Cruikshand, 1987). For procedures being carried out in practice a basic distinction is often made between public, private, conventional and alternative procedures. "Private" procedures are those in which conflicts without any public interest or without the involvement of public institutions are settled in a hierarchical way (for example arbitration negotiations to settle labour disputes). Conventional procedures for settling disputes include traditional, administrative, legislative and judicial procedures. They are governed by formal regulations which come into force when particular conditions exist; the regulations/procedural stipulations leave the parties in the dispute little scope to determine independently the course of the procedure and its content.

Further distinctions are made regarding the subject of conflict. For example, the term policy dialogue is used to refer to the process of agreeing upon relevant social and economic basic rules (code of behaviour for specific industries, proposals for legislation or political programmes). In the case of negotiation about legally-binding regulations and standards in the public sector, the terms 'regulatory negotiation' is used and usually shortened to 'reg neg'. Information exchange and joint problem-solving is concerned with creating a common base of knowledge (on which there is consensus). Consensual conflict management procedures mostly deal with concrete planning schemes and construction project.

This is not the place to go into detail on the great range of terms used in the field of alternative conflict management which are becoming increasingly difficult to keep track of. This study is concerned solely with one of these forms: the mediation procedure. In this, as in other forms of ADR, *negotiation* plays a central role, with the difference that here an independent person organises and guides the negotiations. Correspondingly a distinction is made between "unassisted" and "assisted negotiation" depending on whether an independent person is involved or not. *Mediation* is a tool especially used for resolving disputes involving environmental issues. It is the approach which will be under discussion in this paper.

3. Mediation Procedures in the Field of Environmental Conflict

Mediation as an informal *and* formal instrument for managing environmental conflicts was probably first used systematically in Japan,where the consensus-orientated political traditions generally support conciliatory procedures for settling disputes (cf. Tsuru and Weidner, 1989).

However, its career as an alternative instrument for settling environmental disputes, an instrument which is now greatly discussed in Europe too, began in the U.S.A. Here this path was first followed in the mid-seventies in the field of the environment – it had already been used in the sixties to settle community disputes and labour-management disputes. The initiative in the environmental sector was taken by Gerald W. Cormick and Jane E. McCarthy who, with financial backing from the Ford and Rockefeller foundations, successfully carried out a mediation project concerned with settling disputes over the planned construction of dams on the Snoqualmie River in Washington State in 1980 (cf. L. Dembart and R. Kwartler). This procedure, the first explicit effort to mediate an environmental dispute, began in 1973 when Cormick and McCarthy initiated discussion with several parties to a flood-control and land-use planning conflict concerning the river. At the end of 1974 the effort ended up successfully with a written agreement between about a dozen parties involved in the conflict.

This "Pioneer procedure" was followed by others carried out by other people and by the beginning of the eighties the experimental phase was already concluded: mediation was institutionalised and professionalised. Mediation had evolved from being a promising experiment to being a widely accepted public policy option.

In many U.S. states there have been numerous private and semi-private mediation institutes for several years now which offer their mediation services and train mediators. Many of the major non-governmental environmental organisations (such as the World Wildlife Fund, Conservation Foundation) favour this procedure, some of them with great enthusiasm. It has also gained popularity and support in private enterprises and governmental institutions. In view of this and its relative success in practice, it is being called an "economic growth sector" and there is even talk of a "mediation boom".

3.1. *Developments in the U.S.A.*

In the U.S.A., mediation procedures are used at all levels of government (local, regional, state and federal) and in a wide variety of political spheres: for example, in land-use decisions,licensing of stationary sources, infrastructure programmes, law-making procedures, fundamental political declarations, development of codes of behaviour for specific industries regarding particular environmental issues and in the development of general codes of behaviour for dealing with environmental conflicts. A study carried out by Gail Bingham summarised this development as follows:

> Nationally, by the end of 1977, nine environment disputes had been mediated. Another eleven were mediated in 1978, and nineteen more were mediated in 1979. By mid-1984, mediators and facilitators had been employed in over 160 environmental disputes in the United States . . . . In addition, and relatively recently, the practice of environmental dispute resolution has grown beyond the resolution of disputes on a case-by-case basis to the institutionalisation, by statute, of procedures for resolving environmental disputes. Statutes in Massachusetts, Rhode Island, Texas, Virginia and Wisconsin authorise or even require negotiation of disputes over the siting of solid waste or hazardous waste facilities. A statute in Virginia specifies procedures for negotiation and mediation of intergovernmental disputes triggered by annexation proposals . . . .
>
> Environmental dispute resolution techniques have been used to settle, or attempt to settle, a diverse assortment of conflicts. In general, however, the primary issues involved in these cases can be divided into six broad categories: land use, natural resource management and use of public lands, water resources, energy, air quality, and toxics. These categories can be subdivided into side-specific and general policy categories. (Side-specific disputes are over a particular project or plan; others involve disputes over questions of state or national environmental policy, that means general policy categories. Added by H.W,) (Bingham, 1986, pp. xvii, 30).

Cormick/Knaster (1986, p. 7) consider the sharp increase in mediator-assisted negotiations within the development of general consensus regulations to be particularly worthy of emphasis:

> The greatest current expansion in the use of mediated negotiations is in the development of consensus regulations, where parties in conflict are brought together to hammer out regulations that all parties find acceptable. This process has come to be known as regulatory negotiation or "reg neg". The concept of negotiating regulations was originated and developed by Philip Harter and has been used by a number of federal and state agencies. The first three cases involved the negotiation of regulations on non-conformancy penalties for vehicle emissions at the Environmental Protection Agency (EPA), on crew flight and duty time at the Federal Aviation Administration and on the benzene exposure at the Occupational Safety and Health Administration.

In what is probably the most comprehensive study to date of mediation procedures for settling environmental disputes (Bingham, 1986) the balance sheet proves very positive for this procedure: of the 132 cases looked at, in which all parties were trying to negotiate agreement with one another, this aim was achieved in 103 cases (78%). A large number of these agreements have also been implemented.

The generally positive results which Bingham identified in her analysis of case studies going back to 1984 are backed up by more recent studies (cf. Crowfoot and Wondolleck, 1990, with their references for further reading). However, there are also critical voices which claim that the positive effects of mediation are overestimated, failures not sufficiently taken into account, the manipulative character of this kind of procedure is partly overlooked and that generally speaking the positive aspects are pushed to the foreground, because in the meantime a large institutional and financial interest in carrying out these procedures has developed. They also point out that the results are often analysed by the mediators themselves or by scientists from institutions kindly disposed to mediation (cf. Amy, 1990, 1987, 1983 with further bibliographical references). I shall give an outline of the essential arguments for and against mediation. It will of necessity be brief.

Advocates of mediation as a rule focus on its advantages over conventional instruments for implementing environmental policy. They say that mediation procedures lead to fairer, more effective, efficient, rapid results which are acceptable to all parties to the conflict. Susskind and Cruikshank, for example, summarise the results of their own studies and those of others on alternative conflict resolution as follows:

> Based on the research and writing of a great many experts who studied the broad spectrum of disputes, we have identified four characteristics of a good negotiated settlement: fairness, efficiency, wisdom and stability. (1987, p. 21)

A particular advantage of mediation and other alternative methods of conflict resolution, they go on to say, is that, unlike conventional instruments which produce win/lose solutions, they produce all gain solutions.

A particularly important point is considered to be the potential of these procedures to initiate social and political learning processes which are to be welcomed especially from the point of view of democratic theory and societal reform:

Our view of why social reforms often fail is that they are imposed from above . . . . One of the most exciting aspects of consensual approaches to dispute resolution is that once people use them, and find that they work, those people become advocates. The act of participation, and especially the fruits of success, changes their sense of how best to proceed . . . . Unlike most reforms, consensus-building has the great advantage of built-in learning. The reformed become the reformers. We contend that it is precisely this kind of learning that allows a reform to take root and endure. (Susskind and Cruikshank, 1987, p. 246)

Critics of mediation procedures point out a number of weak points using the example of individual case studies: the criticism is as a rule convincing but often applies only to the specific circumstances of the particular case from which deficits inherent to the instrument cannot necessarily be inferred. Nevertheless some inherent deficits are pointed out which evidently call the above-mentioned advantages into question. Douglas J. Amy summarised this fundamental criticism as follows:

At first glance it is difficult to see why anyone would criticise the environmental dispute resolution approach. Why would anyone be against such desirable things as cooperation, communication and win-win solutions. Nevertheless, there are critics of this process and they have three general areas of complaint: first, that many of the EDR (environmental dispute resolution) advocates' criticisms of litigation are exaggerated and inaccurate; second, that EDR may work more to the advantage of business groups than environmentalists; and third, that EDR fosters a distorted understanding of the nature of environmental conflicts. (1990, p. 221).

To substantiate their claims, critics point out that when mediation procedures are used for industrial projects which are rife with conflict they are highly time-consuming and expensive. Even advocates of mediation, such as Gail Bingham, were unable to show that these procedures are more rapid and less expensive:

Perhaps the single most common assertion made about environmental dispute resolution processes . . . is that they are cheaper and faster than litigation. There has been little empirical evidence to support this assertion, however. (1986, p. xxv)

With regard to fairness, critics argue that, contrary to promises, no truly broad participation is made possible but that well organised environmental interests dominate while less powerful groups are consciously or unconsciously left out. In the procedure itself it has not been possible to eliminate the asymmetrical distribution of power in society to such an extent that representatives of environmental interests can negotiate with the same skill and corresponding success as highly-trained representatives of business or the government. Finally a fundamental unresolvable dilemma is pointed out: The major environmental disputes were based on differing (and deep-rooted) values and principles which are fundamentally incompatible. This type of constellation cannot actually accommodate compromise, at least not between the interests of industry and the environment. The following conclusion would have to be drawn from this:

> If many environmental issues are of the nonnegotiable type, then it would follow that a more aggressive and adversarial form of environmental politics should be practised. This vision of environmental politics would embrace litigation, and would celebrate, rather than criticise, the win-lose style of decisions handed down by the courts because this allows for the complete vindication of the environmental position. (Amy, 1990, p. 227)

In view of the current state of research it is probably not possible to make a conclusive, balanced and empirically plausible judgement of the substantial advantages and disadvantages of mediation procedures. Examination of the relevant literature on this instrument of conflict resolution in environmental issues reveals that the advocates, or at least the most prominent ones, generally tend to push the positive aspects of mediation into the foreground or to consider that in the long-term it will be possible to overcome the obvious failures by improving the instrument itself. Those who are fundamentally critical (a clear minority), tend on the other hand to over-emphasise individual failures and to generalise. In contrast to the advocates their direct empirical experience of these procedures seems to be rarer and they have also carried out fewer empirical studies on the subject. Common to both sides is that they make generalised, universal statements about the chances for and problems of mediation procedures for settling environment disputes in highly industrialised societies which are based predominantly on the experience within the political system of the U.S.A. This is bound to be the case, since – with the exception of Japan and possibly Canada – other countries have only had rudimentary experience with this instrument. Finally, it is obvious that critics and advocates alike in the U.S.A. take often fundamentally different attitudes (concepts, theories, ideologies) to the possibilities for ensuring or improving the ecological basis under existing political (capitalist) systems.

In other countries too, the discussion on this has by no means been concluded. In the Federal Republic of Germany, however, where a heated and sometimes very fundamentalist debate has been raging for 20 years, a consensus is gradually being formed amongst political scientists; the opinion developing is that in the course of a "modernisation of the state" a positive relationship between the market economy and the environment will become possible (in the form of an "eco-social market economy") (cf. Böhret, 1992, and the discussion in Kohler-Koch, 1992).

Society is often in practice not willing to wait for the outcome of this theoretical and ideological dispute. The use of mediation in the U.S.A. is constantly on the increase. Even Amy, one of the most severe critics of mediation procedures also points out: "In any case, it is clear that environmental dispute resolution has now gained a foothold in environmental politics and that it is here to stay, in one form or another." (1990, p. 232) and in European countries there is growing interest in all areas of society; a few procedures have already been initiated. In 1992, with a time-lag of some twenty years, mediation has reached in Europe more or less the level in practice it had in the U.S.A. in 1975.

3.2. *Developments in Europe and Japan*

In comparison to the U.S.A., Japan and Canada, mediation procedures are virgin territory in European countries both in terms of theory and practice. Mediation procedures in the "classical" sense, i.e., which follow the basic principles established in the U.S.A. are at best to be found in Germany, Austria, Switzerland and the Netherlands (cf. Fietkau and Weidner, 1992; Knoepfel and Rey, 1990; de Soet, 1988) but even here the phase of implementation of an agreement resulting from a mediation procedure has to our knowledge not yet been reached. However, in the European countries mentioned and a substantial number of others too, other alternative forms of conflict settlement which involve negotiation and bargaining are common practice. This applies both to countries with a centralised structure and those which are organised decentrally particularly on the implementation level of government (e.g., Great Britain, Sweden and France).[3]

3.2.1. *Japan*

Throughout long centuries of isolation from foreign influence, independent, homogeneous forms of behaviour, social roles and a specific legal mentality formed in Japanese society which is very much a society of segregated groups. They still have considerable influence today. (Leubel *et al.*, 1979) Traditionally, conflicts were not settled on the basis of the rights of the individual but primarily under the aspect of what the effect on the community would be. Conflicts were not dealt with on the basis of formally fixed methods for bringing them into the open to be settled but instead, in line with the Japanese desire for harmony, every attempt was made to avoid dispute and find a solution involving both sides giving in. As a rule conflicts within groups would be settled by mediation by other people or institutions.

As early as in the Tokugawa era, traditional conciliation procedures (*naisai* or *atsukai*) were already highly developed. The role of a conciliator or mediator was taken on by people of high repute or social status. Although Japan thoroughly modernised its legal system on the lines of Western examples as part of the Meiji reform (or restoration) of 1868, current Japanese legal practice and mentality is still strongly influenced by tradition: correspondingly there is a wide range of procedures within and outside the legal system which permits disputes to be settled out-of-court. This is not only due to historical reasons. One of the reasons that the possibilities for settling disputes in court, which were actually extended after the Second World War both by the constitution and individual laws, are made such little use of is that they cost a lot of time and money and that public administration itself has been building up a complex system since the fifties which offers the public a service for out-of-court settlement of disputes.

The extremely vehement environmental conflicts in the sixties gave rise to the

Basic Law on the Environment, of which Article 21 laid the foundation for special legislation to define out-of-court procedures. In this the government is assigned the task of establishing procedures for settling environmental disputes in the form of mediation, conciliation and arbitration which also include the possibility of financial compensation. This was put into practice in the form of the "Law on Settlement of Disputes in Connection with Environmental Damage" of 1970 and the "Law on the Establishment of a Committee to Settle Matters of Environmental Damage" of 1972.

Three separate institutions were set up to deal with out-of-court settlement of environmental conflicts:

- *Complaints commissions* at local authority level whose function in many respects corresponds to that of an ombudsman.
- *Prefecture commissions* to settle disputes over environmental damage deal mainly with disputes between private individuals and between private individuals and the administration. The members of the commission are independent. Depending on the case concerned they carry out mediation, arbitration or conciliation procedures.
- *The central commission* for dealing with environmental damages is an independent government office reporting directly to the Prime Minister's Office. It has a semi-court function. It can also carry out mediation, arbitration or conciliation procedures and in addition it is the only institution which can carry out what are known as "verdict procedures" to determine liability or cause.

Essentially there are four different procedures for settling disputes:

- Mediation
- Conciliation
- Arbitration
- Verdict on guilt/liability and cause

In the case of mediation the role of the commission is primarily to act as go-between for the parties in dispute until they can find a mutual basis for agreement. The result is recorded in a memorandum. Mediation is the procedure for conciliatory settlement of disputes which is most deeply rooted in tradition. It is also widely used in civil and public law cases (cf. Sagami, 1989). Apart from a few exceptions the costs of the procedure are borne by the parties involved; they are significantly lower than normal court procedures.

National statistics show that of the different procedures available mediation procedures are clearly favoured. On the whole the procedures for settlement of disputes (and the complaints system) are considered to have very positive effects. They are seen as being relatively flexible and accessible to the public (cf. Upham, 1987). The nature of the procedure means that the measures connected with it are related to the individual case and, as in the majority of cases it is not a question of implementing legal stipulations, the measures are flexible in that they not only take into account the local conditions and technical possibilities but also the economic situation of the source of complaint. It can generally be assumed that economically acceptable compromises will be found. It should be emphasised that due to the conciliatory dispute settlement systems the negotiating position of the parties involved is stronger, particularly in cases where a court case would be either difficult or impossible, since the legal barriers which would otherwise exist are to a great extent eliminated. The different procedures for settlement of disputes also provide the parties involved with the choice of yet another instrument for voicing their interests; the option of making recourse to other ways of implementing their interests still remains open.[4]

3.2.2. *The Federal Republic of Germany*

Mediation procedures are an extremely rare element in the environmental policy of the Federal Republic of Germany; other types of procedure are far more common. At present, in virtually all areas of the environment, pre-negotiation or alternative procedures take place prior to formal procedures and work on an extended definition of the term co-operation, taking into account the interests of the environment and the local people. Most of these cases are related to waste disposal. To name just a few examples where new forms of conflict settlement are being tried out: the search for a site for a household waste landfill in the municipal district of Hildesheim and for a landfill for hazardous waste in the county of Arnsberg; risk assessment, safety measures, and clean-up of a populated contaminated area in the cities of Wuppertal, Essen and Hamburg; safety measures and clean-up at a landfill in the village of Vorketzin in Brandenburg, clean-up of a sludge landfill in the city of Bielefeld, modification of a waste incineration plant in Bielefeld-Herford, setting up of a traffic forum to work out a long-term traffic and transport concept for the city of Heidelberg, development of a hazardous waste plan for the federal state of Lower Saxony, setting up of a working group on waste management in the county of Osnabrück, planning process for a landfill for sludge dredged from the docks in Hamburg; clean-up measures for sources of pollution and installation of plant for environmental protection in companies (Hamburg, the village of Seelze), examination of the impact on the environment of producing a pesticide at the Hoechst company and a dialogue (co-ordinated by the WZB) to assess the consequences of cultivated crops which have an in-built resistance to herbicides induced by genetic engineering (cf. van den Dale, 1991).

In the cases mentioned, which take place in the context of regional planning, environmental impact studies, official approval of plans, licensing procedures and other formal procedures, discussions take place in small working groups, and moderators, people specially appointed or consultants are given the task of providing information and creating a compromise within the group. In the Federal Republic of Germany there are only two real mediation procedures which measure up to the standard developed in the U.S.A., that is to say which deal with cases of conflict which can be classed as major both in terms of political-social importance and the project at stake and which involve an independent mediator. One is the mediation procedure in the village of Münchehagen in Lower Saxony and the other in the district of Neuss in North Rhine-Westphalia. Both are to do with questions of waste management.

3.2.2.1. *Mediation Procedure on the Waste Landfill in Münchehagen.* The background to the mediation procedure in Münchehagen (Lower Saxony) was years of controversy over the Münchehagen hazardous waste landfill which centred around suspicion of illegal depositing of waste, contamination of soil and water and health risks. In the course of events, which was rife with scandals, a court declared the landfill to be in contravention of the law, a special commission was set up in the criminal investigation department and a committee of investigation was also set up in the Parliament of the *Land* of Lower Saxony. The parties in the dispute – representatives of the *Land*, various town councils, local authorities and county authorities and several pressure groups – were deeply split on the issue.In this situation Meinfried Striegnitz, a member of the Loccum Evangelical Academy, situated near the landfill, became involved. He organised and chaired a number of discussions with experts which nearly all the parties took part in. This essentially positive experience led to the idea of setting up a broader-based mediation procedure. The idea became reality when the Münchehagen Commission was established at the end of 1990 and Meinfried Striegnitz was appointed as official mediator by the Environment Minister of Lower Saxony.

The aim of the procedure is to get the disputing parties to agree on a clean-up method. The role of the mediator is primarily to sound out whether there is any scope for manoeuvre, identify conflicts early on, and maintain an uninterrupted dialogue. If controversial issues cannot be settled under the guidance of the mediator the parties to the conflict, before instigating a court case, have the possibility of calling upon an independent arbitrator. To date this has not been necessary.

The mediation procedure has not yet been concluded; the experience up till now has, however – particularly in view of its pioneering function – been very promising. For example, the procedure has managed to get the parties which were previously in conflict to work together in a constructive, objective manner. Prejudices have been overcome and an open atmosphere for dialogue has been created which proved very fruitful for the development of trust. For some problems which had previously been the subject of great controversy it was possible to find solutions which all parties accepted. In particular it was possible to eliminate the lack of information which had been bemoaned previously. The initial results of a review of the mediation procedure have shown that virtually all parties involved evaluate the procedure positively. Those questioned stressed the positive function of the mediator in improving the general and particular communication between the parties in dispute and in furthering the process of problem-solving. Recently, in August 1992, an agreement was reached for the first time by all parties to the conflict on general clean-up objectives, safety measures and criteria for evaluating them (cf. Striegnitz, 1990; Pfingsten, 1992). The aim is to come to an agreement by the end of 1992 on an overall concept for the clean-up of the landfill.

The second major mediation procedure is going on in the district of Neuss and is concerned with the waste management plan for the district. Below is a detailed report on the procedure which has been the subject of a parallel academic investigation by a mediation project team (of which the author is co-director) within the WZB (mediation project).

3.2.2.2. *The Mediation Procedure on the Waste Management Plan in the District of Neuss.* Since 1990, social scientists at the WZB have been carrying out a broadly based research project on mediation procedures in environmental policy using the problem of waste management as an example (cf. Fietkau and Weidner, 1992). The central subject of investigation is a mediation procedure concerned with the local authority's waste management programme which is being carried out in conjunction with the district of Neuss (North Rhine-Westfalia). In addition to this mediation procedure, which is at the centre of the analysis, experiences with other mediation procedures in Germany and abroad are evaluated.

At the centre of the project are the following questions to be researched:

- How do mediation procedures differ in political and sociological terms from other procedures for developing environmental standards?
- Can mediation procedures be successfully used in the Federal Republic of Germany to solve environmental problems?
- How can they be integrated into the political and administrative system of the Federal Republic of Germany?
- What elements in the mediation procedure, which can be identified and described in sociological terms, can be said to further or impede its success?
- How do the public and the institutions involved react to the procedure?

The examination of these research questions is being carried out in an inter-disciplinary manner and includes international experience and the experience of people involved in the practical work of the relevant fields (cf. Pfingsten and Fietkau, 1992). In academic terms this research represents the exploration of virgin territory.

To date there are no results of evaluative research by social scientists in Germany on mediation procedures. Due to the differences in the legal framework and political tradition, the most advanced Anglo-American research can be transposed only to a limited extent to the German situation. Apart from that it also has shortcomings in that it primarily analyses mediation procedures in retrospect. We are hoping that a parallel research project, like the model project being carried out at present in Neuss by the WZB research team, will make it possible to produce results which are not filtered through the memories of those involved and which, due to the directness of the observations, do not have to be content with summarised versions which is necessarily the case when documents (minutes) are analysed. These analyses of procedure, which are very time-consuming for the researchers, are currently being carried out. Since it is important not to interfere in the on-going process, it will not be possible to report on them until a later date.

In September 1991 the local council of the district of Neuss approved a waste management programme. It is intended to provide "reliable disposal" as required by law and to put into practice the priority given in principle to avoidance and recycling over "disposing" of waste. The remaining refuse is to be incinerated and the incineration residue deposited in a landfill. Various aspects of this programme were discussed by the local people, some of them were the subject of great controversy. The issues included:

- the full use of all possibilities for avoiding and reducing household waste and waste from trade and commerce.
- the use of appropriate methods for separate collection and sorting of recyclable waste fractions
- the increase in the proportion of waste generated in the district which is recycled
- the necessity and capacity of a waste incineration plant
- the siting of new plants for sorting, recycling, composting, treatment, incineration and landfilling waste

The WZB, with the co-operation of the local authority, initiated a mediation procedure for the waste management programme.

Professor Dr. Georges Fülgraff (formerly president of the Federal Agency of Health and member of the expert committee advising the government on environmental issues) agreed to act as mediator. In the long and tedious search for a mediator we focused on the following criteria which we thought the post required: impartiality, competence in the subject matter and in dealing with people, experience in politics and overall reputation.

After a period of several months of communication between the WZB research team and all groups with potential interest in the case, there was extensive approval in the district both for the procedure itself and the choice of mediator. To support the on-site organisation of the procedure the WZB set up a mediation office in the district.

After lengthy negotiations and numerous bilateral and multilateral preliminary discussions between the mediator and potential participants in the procedure, the first communal mediation meeting took place in the town of Grevenbroich in March 1992. Over 60 people took part, representing some 30 organisations and institutions from the field of politics, the administration, nature conservation organisations, churches, pressure groups and industry. This meeting constituted the beginning of the mediation procedure. A second meeting in Neuss in May 1992 produced the first concrete results: a moratorium was agreed: no binding stipulations were to be made, particularly as a result of the action of the administration, until a series of expert reports were available which would enable the discussion to continue on a sound, objective basis. This includes the intention to have consultants trusted by the pressure groups comment on the existing waste management programme.

A Short Chronology of the Mediation Procedure in the County of Neuss

December 1990
The administrator in charge of the environment in the county of Neuss, Dr. R. Fonteyn, indicated that he is interested in carrying out a mediation procedure in conjunction with the WZB to deal with the waste problem in Neuss.

February 1991
The WZB research team and the local council discussed the mediation procedure and agreed on a joint launch of the project.

June 1991
We presented our mediation project at the public "waste forum" in Neuss and proposed our candidate Professor G. Fülgraff as mediator. Both the project and the choice of mediator met with approval.

August 1991
The mediator took up his post.

September 1991
In September 1991 the Neuss county council passed a waste management programme. It is intended to provide "reliable disposal" as required by law and to put into practice the priority given in principle to avoidance and recycling over "disposing" of waste. The remaining refuse is to be incinerated and the incineration residue deposited in a landfill.

October 1991
A mediation office was set up in the county of Neuss.

13 November 1991
A co-operation agreement on the mediation procedure was signed by the Neuss local authority and the WZB research team.

November 1991 – January 1992
Preparatory discussions in advance of the first mediation meeting: A total of 21 preliminary discussions were organised by the mediator and the WZB mediation project team with all potential participants. In general there was approval for the procedure and willingness to participate.

28 March 1992
On 28 March 1992 the first joint mediation meeting took place in Grevenbroich. Over 60 people took part, representing some 30 organisations and institutions from the field of politics, the administration, nature conservation organisations, pressure groups, churches and industry. This meeting constituted the beginning of the mediation procedure.

The first concrete step was the announcement of the intention to have consultants trusted by the pressure groups comment on the existing waste management programme. It was agreed that no further measures should be taken until the expert report had been completed and discussed.

8 May 1992
Individual questions concerning the expert report were discussed by the mediator and the pressure groups. They developed a catalogue of questions. The consultants to prepare the report were selected.

26 May 1992
2nd large mediation meeting. The proposals of the administration on the appointment of consultants are discussed by the 40 participants. Agreement was reached.

3 July 1992
Meeting held between the district administration, the pressure groups and the consultants to discuss a report on environmental implications for the area. A report was commissioned to describe the effect on the environment and the three possible sites for the waste incineration plant currently under discussion.

10 July 1992
Meeting between the district administration, pressure groups and consultants to discuss a health study. A report was commissioned to describe the current state of health of the population at the three possible sites for the waste incineration plant.

October 1992
The third large mediation meeting will take place.

Below is a description of some of the experiences we gained during the preparation process for the procedure in Neuss.[5]

Determining the Group of Participants. At the beginning of the mediation procedure the question arises as to who should take part in the procedure and who should decide this. Should it be everyone affected, everyone interested or all those who can legally or politically intervene in the planning process? In preparing the Neuss mediation procedure we talked to public administrators at different levels, parties, pressure groups, trade unions, employers associations, companies and churches. We invited everyone we knew to be interested to the first meeting. Some 60 people from 30 organisations came. In such a large group it is very difficult to have a direct exchange of opinion. If everyone has something to say about a particular problem there will soon be a severe time problem. Spontaneity therefore has to be dampened by centralised running of the meeting. We suspect that as the mediation procedure continues there will be a reduction in the number of participants. The opportunities for discussion can, however, be ensured by holding meetings in small groups in addition to the large mediation meetings in order to deal with individual points on behalf of and in preparation for the mediation meeting.

Equal Treatment. Equal treatment of everyone involved in the mediation procedure is one of the basic conditions for mediation. This approach, however, is incompatible with political and legal reality. The county administration as planning authority responsible for the waste management programme and the president of the regional government as licensing authority have by virtue of these functions special roles (*primus inter pares*). This has consequences for the legal design of a mediation procedure (cf. Holznagel, 1990) and also causes problems in the social and political preparation and implementation of the mediation procedure.

Since the county administration drew up the waste management programme and organised the political decision-making process connected with it, it automatically had an ambivalent attitude to the mediation procedure. On the one hand the mediation procedure could disturb the overall progress and yet on the other hand it is seen as providing an opportunity to deal with conflicts in a well-ordered and efficient manner. The administration naturally has an interest in having its ideas accepted in the mediation procedure. It thus at one and the same time represents the interests of and administrates the political process. As main participant it also becomes one of the main partners for dialogue in the preparation of the mediation procedure. The close involvement of the administration in the mediation procedure arouses the suspicion of the other participants that mediation could simply be an instrument to help the administration enforce its will. This is certainly true to some extent and yet in this the administration does not differ from the other participants. Nevertheless, this constellation does confront the mediator with the problem of having to work closely with the administration, because otherwise a mediation procedure would hardly be possible, whilst at the same time not allowing himself to become involved in the interests of the administration and, what is more, making this clear to the outside world. Credibility for the impartiality of the mediator was considerably facilitated by the fact that this work, his office in Neuss and certain other things were financed by the WZB which receives funding for the mediation project from the Ministry of Research and Technology. Nevertheless, the pressure groups and representatives of the Green Party were sceptical at the beginning as to whether the mediator and the WZB did not have undeclared interests behind the veil of impartiality being presented to the public. This suspicion made them sensitive to individual actions taken by the mediator and the WZB. We believe, however, that the distrust disappeared over the course of the procedure.

Parallel Research. One peculiarity of the Neuss procedure was the parallel academic research which went with it and, hand in hand with this, the fact that it was initiated and financed by a research body. This prevented a "normal" consultant/commissioning office relationship between the mediator and the county administration. This has the advantage of making it easier for the mediator to implement and display his impartiality. However, it also has a disadvantage in that the impression could be gained that the mediator had not been appointed but had imposed himself on the procedure. The "normal" aim of a mediator of earning money through his work has been overlaid by the seemingly diffuse research interest. The personal interest of the mediator and the WZB make it seem that both these parties are difficult to keep a check on-going as far as producing uncertainty as to what will be written in the academic documentation about those involved and if this could not prove disadvantageous to them.

Conditions for Initiating a Mediation Procedure

On the basis of the experience of initiating a mediation procedure in Neuss as well as the experiences of others working in this field, a number of conditions (which make no claim to being exhaustive or systematic) can be named which are important when launching a mediation procedure:

- All concerned, or at any rate the central participants, should have the expectation that negotiations will swing the outcome of the problem-solving mission in their favour and/or that the problem-solving process itself will be more advantageous to them than a conventional (legal) procedure.
- All potential participants must be prepared for the procedure. A minimum of consensus must be achieved in which the parties agree on forms of dealing with each other.

- For the preparatory work a person must be appointed to whom all participants can voice their opinions and who enjoys the trust of all parties. To ensure communication amongst the participants and to deal with organisational matters and the problem in hand the mediator must set up an office "on the spot".
- A basic outline of the problem must be presented, whereby it must be ensured that the position of all the participants remains flexible.
- The groups and institutions involved must appoint individuals to represent their interests (if possible with continuity) throughout the procedure.
- It is important to clarify the cost aspect of the procedure. The type of financing must avoid inherently favouring any particular solution and any doubts in this direction must be eliminated.
- The relationship between the mediation procedure and the legally prescribed procedure for the case in hand must be clarified as far as possible.
- Relationships with the press and the public for the duration of the procedure must be clarified.
- The potential participants in the procedure must agree in advance on a mediator to run the actual procedure.

The first results from our analyses of the mediation process are now available – the mediation meetings are recorded on video and audio tape and a questionnaire is circulated to all participants on the day of each meeting. For practical reasons and reasons of methodology it is not possible to publish these results in detail whilst the procedure is still running. This would be an intervention on the part of the observing scientist and could interfere in the dynamics of the procedure. The results listed below have thus been generalised; they are based on evaluations of the questionnaire which was distributed at the second mediation meeting:[6]

- Most of the participants in the procedure have no previous experience with mediation or similar procedures. This is unfamiliar ground for them.
- The participants have a very positive opinion of the mediator. He is virtually unanimously perceived as being competent, fair and sure of his work.
- Almost half the participants consider a solution approved of by all parties to be feasible. The other participants are either uncertain or sceptical about this.
- Approximately one third of the participants said that the joint mediation meeting had changed their view of the problem and/or the other parties to the procedure.
- Approximately half of the participants believe that the proposed waste management plan could have negative effects. Possible risks are perceived to be health problems for local people, the political climate, the ecological situation and the quality of life in the county.

- The majority thought that the waste management plan would boost the economy of the county but would not be so useful to other areas of life (health, ecology, politics, quality of life).
- Almost half of those involved in the procedure believe there are alternatives which would make it possible to realise the economic benefit of the waste management plan in a different way.
- When asked to place the aims of the procedure in order of importance, the average list of priorities was as follows:
    1. a solution which would last;
    2. a fair solution;
    3. an unequivocal result;
    4. a solution approved by the general public;
    5. a solution which can be easily implemented;
    6. a rapid solution.

To conclude; to date there has been little experience with mediation procedures in the field of environmental policy in the Federal Republic of Germany. It is nevertheless possible to make use of the foreign concepts, particularly those developed in the United States, Japan and Canada.

Mediation procedures are the subject of increased attention by academics. Particularly jurists have begun to analyse and try to develop standards for this procedure. In the meantime social scientists are also focusing more attention on the mediation procedure. The academic observations made in the Federal Republic of Germany are, however, often of a purely speculative nature; it is obvious that they lack concrete experience. Whether and how it will be possible to successfully carry out mediation procedures within the legal framework and political tradition in the Federal Republic of Germany is still an open question. To answer it empirical hard labour is required; the institutional and theoretical "superstructure" can then be built up.

4. Mediation as a Policy Instrument

As I have described in the previous chapter, in some countries (U.S.A., Japan, Canada) mediation is already common practice in the environmental arena. In other countries it is still something of an exception but is being increasingly used.

To my knowledge, mediation in the field of environmental policy was a "social invention". It was not incorporated into the arsenal of environmental policy instruments as a compulsory or optional standard procedure but owes its existence to the initiative of some social groups. In the meantime (particularly in Japan and the U.S.A.) mediation has also become a formal governmental instrument and has to varying degrees been standardised. In individual cases the law requires that it be used before recourse is made to other procedures for dispute resolution and decision-making (U.S.A.). However, not even this has made it an exclusively governmental instrument. Governmentally unregulated mediation procedures still take place when social groups take the independent decision to try out this instrument.

Can a "hybrid" of this kind be counted as an environmental *policy* instrument? My answer is yes, for reasons I shall set out below. This explanation is not intended to be a further contribution to the discussion on classification of environmental policy instruments. That is an admirable exercise which is proving to be infinite and which many social science students have devoted time to.

Classifications in political analysis should not be an end in themselves (cf. Mayntz, 1980, 1983; Kaufmann and Rosewitz, 1983). For political analyses they can be very helpful, since questions hypotheses and theorems which have already been developed for and possibly empirically tested on existing classes/types can be applied to new instruments which have still to be researched – such as mediation. This facilitates the design of theory-driven empirical studies.

First, I shall reiterate briefly what is meant by mediation. To do this I shall use definitions developed in the U.S.A. because, to my knowledge, the mediation procedure has been given most careful consideration there by social scientists (including economists and legal experts).

Gerald W. Cormick (1980), who pioneered the use of the mediation process in environment conflicts, defines mediation as:

> A voluntary process in which those involved in a dispute jointly explore and reconcile their differences. The mediator has no authority to impose a settlement. His or her strength lies in the ability to assist the parties in resolving their own differences. The mediated dispute is settled when the parties themselves reach what they consider to be a workable solution.

Another definition of mediation (MacDonnell, 1988, pp. 12 ff) which includes definitions by other authors, states that:

> Mediation introduces an outside neutral into the settlement process to act as a facilitator. Stulberg has provided the following explanation of mediation: "The mediation process can be characterised as follows: It is (1) a non-compulsory procedure in which (2) an impartial, neutral party is invited or accepted by (3) parties to a dispute to help them (4) identify issues of mutual concern and (5) design solutions to these issues (6) which are acceptable to the parties." As with negotiation the only rules or structures that apply are those imposed by the parties themselves. No objectively definitive norms or principles are assumed to control the outcome. As Fuller suggests, it is the settlement itself that creates the norm. And, like negotiation, the settlement requires the mutual agreement of the parties.

Gerald W. Cormick (1980, p. 27) in his definition quoted above lists a series of criteria which are essential for a true mediation procedure:

> (1) The parties cannot be required to negotiate or cannot be unduly coerced to agree to any particular settlement of their differences. Indeed, unless they are willing to enter into the process with some intent to reach an accommodation of their differences, the mediation effort is not likely to be viable.

(2) There will be a joint or face-to-face exploration of the issues, that is mediation must be seen as an adjunct to the negotiation process. (3) The mediator supports and facilitates the negotiation-mediation process by improving communications, serving as an interpreter, arranging meetings, suggesting alternatives, helping to draft language, assisting in maintaining communication with those not "at the table", and so forth. Whereas in labour-management disputes the mediator typically enters a dispute to revive lagging or severed negotiations, in environmental disputes the mediator usually serves a primary function in establishing a negotiating relationship. (4) Any agreement reached is the creature of the parties and must be deemed viable and acceptable by them. The mediator is not party to the agreement.

Seen in terms of negotiating, a central device in political theory and even more so in policy analysis, mediation can be defined as negotiation with the assistance of a trusted, independent and impartial third party whereby negotiation itself is defined (Iklé, 1972, p. 117) as:

a form of interaction through which individuals, organisations and governments explicitly try to arrange (or pretend to do so) a new combination of some of their common and conflicting interests.

From environmental policy studies we know that different scale negotiations take place on in environmental policy processes on all levels, formal and informal, at programme development stage and implementation stage and that they are used in conjunction with virtually all environmental policy instruments (even the so-called "command control order" instruments). Unlike these established forms of negotiation which may be legally legitimated or, as often is the case, take place in the shadow of the law, mediation not only includes a third party as guardian of the procedure but negotiation here has to fulfil some basic prerequisites specific to the form of mediated negotiations. These include the following minimum conditions (cf. Cormick, 1980, p. 28):

1. There must be a recognition by all parties of the necessity of other parties participating in the process as co-equals; that means some level of partnership between the parties has to be achieved.
2. Each of the parties involved must have sufficient power or influence for sanctioning other parties' abilities to take unilateral action.
3. Participants should be able to commit themselves and their constituencies to implementing agreements reached in the negotiation process.
4. Participants must have some sense of urgency with respect to settling the dispute.

It is one of the central responsibilities of the mediator to ensure that these basic conditions are fulfilled both before and during the procedure.

It is becoming clear that the mediator would have to possess almost superhuman qualities and skills in order to be able to fulfil the role assigned to him in the mediation procedure. Although the ideal person will seldom be found, practice has nevertheless shown that there are sufficient people with a natural leaning to the job who at least adequately fulfil the requirements. In the U.S.A. institutions grew up in the seventies which run systematic training schemes.[7]

I shall not go in detail into the methodological and practical aspects of designing and implementing mediation procedures nor into the skills of the mediator. The following brief list of criteria is intended to give a general idea:

- Since participation in mediation procedures is essentially voluntary, the mediator, unless he has been called in by one of the parties to the conflict into a set-up they themselves have created, must in advance inform and persuade potential participants.[8]
- The exploration of the issues is carried out jointly by the participating parties being supported by the mediator.
- The mediator has no authority to impose a settlement on the participants.
- The mediator facilitates the negotiation process by assisting the parties in reaching a resolution acceptable to them.
- The mediator shares the responsibility for ensuring that any agreement reached represents a workable (feasible) solution. This means that he has an advisory function for participants who have little or no knowledge and practical experience of the issue at hand (particularly highly complicated ones).

So, is mediation, as defined above, a policy instrument and if it is to what existing categories can it be assigned?

Unlike the relatively recent mediation procedure the term *policy instrument*, well established in political science, has no precise definition. It is a frequently used yet extraordinarily diffuse term. Many political science dictionaries do not even list it separately (nor the term *instrument*) although they often bandy the word around in other listings (such as programme, implementation, decision-making, policy analysis) or use other words to circumscribe it (e.g., political means, methods, devices). It is therefore no wonder that elements of politics which are otherwise very distinct from one another are sometimes referred to as instruments: for instance, some refer to a political programme as an instrument, whilst others would only call the elements or certain elements of a programme instruments.

In German the semantic situation is even more hair-raising than in English, since the German word *Politik* carries the meanings polity – politics – policy – and it is impossible to make a linguistic distinction. German political scientists have started to try to be more precise by circumscribing it and using political structures/institutions for polity, political process for politics, and political substance/political field of action for policy. Policy thus includes the substantial and procedural aspects of political programmes designed to give shape to social relationships; this area includes what is seen as the state's responsibilities for dealing with problems and the strategies and instruments used for this purpose.

In everyday usage *instruments* are appliances and tools (Latin: *instrumentum* –

equipment), i.e. means which are used to carry out a particular task. It would thus seem logical to place them in an aims/means context which is what policy researchers do, in that they define political instruments – i.e. instruments to achieve a purpose or goal – which are needed for political action (to be more precise are used by the representatives of the system of government) as an important element in policy programmes, whereby the programme can be an analytical construction (ideal programme) of the policy researcher or an (expanded/modified) real programme.

Policy analysts see policies or *policy programmes* as a phase in the political process (policy-making, policy cycle), which according to Wildavsky, for example, contains the following elements: perception and selection of problems, articulation and definition of problems, formulation/design of programmes, conclusions/ decision-making, implementation, feedback, evaluation of the effects of the programme, termination of policy-making.

Literature on the subject has no uniform definition for the term *programme* nor is a clear distinction always made between programmes, policies and instruments. A proposal influenced by implementation research for which there is a broad consensus, at least in the Federal Republic of Germany (Kaufmann and Rosewitz, 1983, pp. 34 ff), uses the term "policy" for a perceived group of connected political measures which refer to a particular area of politics, i.e. an institutionalised sector of state policy-making, for example economic policy, financial policy, research policy, social policy.

The term programme, which is usually used in the U.S.A. to designate practical political initiatives, refers to the output from the programme design process and thus the input into the implementation process. A programme as a rule contains a number of inter-related decisions to enable a programme objective to be attained. Ideally it contains the following elements: the problems selected for treatment, the objectives the programme aims to achieve, assumptions about intended effects, and details of the institutions entrusted with carrying out the programme (cf. W. Jann, 1981, p. 49). The policy instruments are as a rule also part of the implementation part of the programme; they serve to implement and enforce the programme objectives.

The instruments chosen (and correspondingly the programmes to which they belong) can be classified into specific types (cf. Kaufmann and Rosewitz, 1983; Howlett, 1991). Foe example, it is common to differentiate between hierarchical and authoritarian instruments (such as decrees and bans) and information and persuasion-oriented instruments, financial and procedural instruments. A *procedural* instrument (or procedural control) serves to create negotiating systems or to institutionalise rules which are intended to reconcile the interests of all parties affected by programmes.

In reality there are often combinations of different instruments within one programme. Policy instruments are means of (direct and indirect) *political control* to fulfil public responsibilities. As such they are also closely connected to the broad discussion on the capacities, capabilities, wills and skills of the governmental system to control other (social, economic, etc.) systems.

Howlett (1991, p. 2) provides the following comprehensive definition:

> 'Policy instruments' is the generic term provided to encompass the myriad of techniques at the disposal of governments to implement their public policy objectives. Sometimes referred to as 'governing instruments' or 'tools of government', these techniques range in complexity and age, although most are well known to students and practitioners of public administration.[9]

Dahl and Lindblom had argued as early as 1953 that the number of alternative policy instruments was virtually infinite. They therefore drew up a classification based on the idea that most policy instruments lay between the poles of five different continua of government choice. In the first continuum, for example, the instruments are classified according to the degree of involvement of private and public companies/institutions. The following table provides an overview.

Table. Dahl and Lindblom's Continua of Instrument Choice

Continuum 1. Nature of Instrument Ownership: *Government to Private Government-Part Government-Joint Govt-Private-Private*
Continuum 2. Nature of Government Influence: *Compulsion to Persuasion Compulsion-Arbitration-Mediation-Conciliation-Information*
Continuum 3. Nature of Government Control: *Direct to Indirect Nationalisation-Licensing-Taxes/Subsidies-Macro-manipulation*
Continuum 4. Nature of Instrument Membership: *Voluntary to Compulsory Private Clubs-Compulsory Membership Organisations-Nation-State*
Continuum 5. Nature of Instrument Autonomy: *Full to None Autonomous Agencies-Semi-Autonomous Agencies-Bureaucratic Agencies*

Quoted from Howlett 1991: 6

The table above shows clearly that mediation is seen to be a policy instrument and is assigned a position midway between "compulsion" and "persuasion".

Now that I have quoted political science literature to prove that mediation is a policy instrument, I can characterise environmental policy instruments as means used or legitimated by representatives of the government system for transforming environmental policy programme aims into environmentally-conscious behaviour. A central aspect of this is that the instruments for attaining goals must involve (e.g., initiating, influencing, stopping) human behaviour; this requirement distinguishes policy from, e.g., technology. Goals without means/instruments are merely a statement of intention.

Mediation procedures can be considered to be environmental policy instruments since they are used or offered by representatives of the governmental system or because the representatives of public institutions participate in them. This is the case invirtually all major mediation procedures on environmental issues. According to this definition it is irrelevant to the classification of an instrument as a policy instrument whether it was initiated by the state, such as the classic "command and control instruments". Moreover, as I have already pointed out, mediation procedures are considered in some countries to be "state approved" instruments, in some cases they have explicitly been incorporated into environmental programmes.

In the context of the most common classifications of policy instruments, mediation can be said to be a policy instrument which has a great deal in common with persuasion and procedural control if the criterion is the medium of control in which the instruments are embedded (authority, persuasion, coercion, incentives). It also belongs to the *negotiation* category.

One of the purposes of categorising and classifying policy instruments is that it helps to understand the factors influencing the choice of policy instruments and to forecast operation and effects. For the field of environmental policy in industrialised countries the accepted academic wisdom is that conventional, especially coercive, instruments often meet with fierce resistance from the affected parties, bring about undesirable side effects and are less than optimal from the point of view of efficiency and effectiveness. This seems to be the main reason why environmental policy researchers show strong preferences for indirect control instruments which leave the parties concerned a greater degree of flexibility in how they adapt to circumstances. This perception in environmental policy studies corresponds to the general opinion of German political scientists working on state control (cf. Scharpf, 1992; Böhret, 1992; Jänicke, 1986).

Modern research into policy studies and state theory does not reject conventional instruments – they have proved very successful in certain areas of environmental policy in the Federal Republic of Germany, for example, as well as in other countries: the act on unleaded petrol, regulation on industrial furnaces and regulation on exhaust emissions from motor cars, to mention just a few cases.[10] It does, however, believe that under the democratic, pluralistic circumstances of complex industrialised societies "softer" forms of control, which are mostly based on negotiation, are more appropriate for managing problems which affect the whole of society and which cannot be solved through "the evolution of society" or by unleashing market forces. Current opinion in German political science, which has taken notice of British and American developments, considers that appropriate forms of control in today's society are contextual, informational and procedural control as well as consensual and dialogue-oriented arrangements.[11] The barriers to greater use of policy instruments based on these central forms are tremendously high in Germany, as will be shown in the following chapter.

5. Ideological and Practical Obstacles to Flexible, Negotiation-based Policy Instruments in Germany

The systematic abuse of non-renewable natural resources, a perceptible rise in environmental pollution, signs of an imminent global environmental disaster, militant tendencies within the environment movement, a growing potential for ecological protest and successful blocking of public and private projects have, particularly in the Federal Republic of Germany, revived the discussion on the question of whether in advanced industrialised societies the state is at all capable of taking appropriate action to counteract trends which are causing damage to the environment. The papers given at the plenary session of the 18th Congress of the German Association of Political Scientists in 1991 were concerned with the question "To what extent is the state at the end of the 20th century able to act" and within this discussion the subject of the environment occupied a prominent position (cf. Böhret, 1992). Essentially the outcome was the diagnosis that there had been an extensive failure on the part of the state which was still labouring under the illusion of being responsible for everything and exerting outdated forms of control. The therapy suggested was that the inadequate, hierarchical, "imperative" forms of control should be replaced by negotiation and mediation procedures.

"New instruments which are flexible and promote co-operation and consensus are what the state needs" was the general tone to be heard throughout the debate on general political theory, which was stimulated by a good deal of input from environmental policy research. To date there has been no response from the state environmental policy-makers: officially, at least, they are still clinging to their established set of dogmas. However, for some time now the increase in unproductive conflicts has increased the willingness to experiment with new methods where negotiations take place on a partner-to-partner level.

There is nothing in Germany's programme on environmental policy to prevent the use of co-operative forms of dealing with conflict. On the contrary: in addition to the polluter-pays-principle and the principle of prevention rather than cure, the principle of co-operation is expressly mentioned as a guideline for the stance to be taken by government offices. However, since it is churned out with boring regularity in official speeches on environmental policy, its essentially progressive spirit is hardly noticed anymore. Formulating the trio of principles in the opposite sense to that intended serves as a reminder of their high intentions: *to be avoided* are subsequent clean-up, nuisance to those not involved or even to those affected and a conflictual stance of the various agents involved. The things described in the programme as undesirable are nearer to the reality of environmental policy than those listed as undesirable (cf. Weidner, 1991).

A more widespread use of solution-seeking through negotiation is thwarted by the problem that in traditional German state theory the idea of a hierarchical relationship between state and society is dominant, the logical consequence of which is that the state is seen as being responsible for virtually all areas of society and as being at the head of the hierarchy. Official environmental policy has essentially followed this state-centered philosophy. In all major (and many minor) areas of environmental protection, state institutions have taken extensive design and control responsibilities (and with them the obligation to draw up regulations) upon themselves. In the field of waste management, which is particularly fraught with conflict, legislators have even – without pressure to do so – decided in favour of a state monopoly on disposal (and therefore virtually a ban on private disposal), which in effect means that the public sector has taken charge of setting up structures to ensure reliable disposal (cf. Hoppe and Beckmann, 1990, p. 17).

The very lively and pluralistic debate on political and state theory, which has been going on for several years and which strongly recommends participatory, non-hierarchical procedures for creating consensus and facilitating decision-making (cf. Scharpf, 1992; Willke, 1991), has had no perceptible influence; environmental regulations still positively exude the spirit of authority and a hierarchical interpretation of the role of the state, which critical observers see as a reflection of "omnipotence fantasies". Correspondingly, conventional environmental regulations still make up the majority of the state's range of standards – a trend which is virtually unaffected by changes of government or statements to the contrary on policy or programmes (cf. Weidner, 1991). Both political scientists and environmental economists alike have not minced words in criticising this. The president of the *Institut für Weltwirtschaft* in Kiel, H. Siebert, complains of a "lack of courage to use the instruments of strict market economics in environmental policy" (Siebert, 1989). In a report for the Ministry of the Environment, the economics experts K.-H. Hansmeyer (for many years president of the expert committee which advises the government on environmental questions – *Sachverständigenrat für Umweltfragen*) and H.K. Schneider complained of the shrinking scope available for flexible environmental policy instruments, a fact which they attribute to the fact that politicians and civil servants believe that the answer to all environment problems lies solely in perfecting their range of regulations (Hansmeyer and Schneider, 1990). The jurist E.H. Ritter describes the same state of affairs in even more drastic terms saying it is about as progressive and democratic as a Prussian army general and claiming that it has reached its limits (Ritter, 1987, p. 937).

The way the state of the environment is developing would indeed justify such harsh judgements since, although Germany has seen two decades of systematic environmental policy-making (based on carefully worked out legal instruments and institutional arrangements), the results as whole can only be deemed modest, despite undisputed success in some individual areas (cf. Umweltbundesamt, 1992). As a result the imminent environmental catastrophes, and linked to them the threat of political crises, are by no means simply the nightmare visions of a handful of people who make a living out of prophesying ecological apocalypses and whose world clock has always pointed to the thirteenth hour. Attention is now being drawn to dangers of this kind in scientific and technical reports prepared by state expert commissions which are otherwise cautious in tone (Deutscher Bundestag, 1990).

Awareness of an impending crisis has generally speaking been made more acute by the realisation that society is causing irreversible processes of destruction in nature which in turn have a ruinous effect on society. It is the possible ecological self-endangerment of the whole of mankind and not simply damage and hazardous developments in isolated areas of the environment that is the main cause for concern which urges counter-measures that will actually change basic structures and a greater willingness for innovation in the political and administrative system. Greater tasks normally demand and justify a change of philosophy and the political courage to experiment with new approaches. The introduction of solutions centred on co-operative negotiation calls for these qualities; it goes far beyond routine changes in policy since, if carried out systematically, it starts to eat away at the idea of an omnipotent state (apparatus). According to Thomas Ellwein, one of the "elder statesmen" of political science in Germany, the main responsibility of the state at the close of the 20th century lies in "legitimating new forms of co-operation between citizens and public authorities in cases where they exist and developing them in areas where they do not yet exist. This would give rise to a fundamental move away from the state model of the 19th century." (Ellwein, 1991, p. 13).

5.1. *A Hierarchical View of the State Impedes Innovative Procedures in Environmental Policy*

Every field of social sciences which has looked at environmental problems has shown a general preference for a fundamental re-shaping of the framework for environmental policy and the instruments for controlling it with a view to achieving greater flexibility, participation and decentralisation. In society itself, "New Social Movements" (c. Roth and Rucht, 1991), of which over the last two decades the environmental movement has had the most influence, have put forward demands for a restructuring of industrial societies to create ecologically-minded economic democracy and an increase in the rights of citizens groups to participate in decision making and have, at least in terms of election campaigns (headword: green parties) successfully represented their interests in the public arena. The willingness of the public to accept effective measures to protect the environment and to participate in them has gradually increased over the years (cf. Dierkes and Fietkau, 1988). In the political and administrative system, however, there has been relatively little happening. The machinery for churning out laws and regulations is working at full speed in order to gain legal control over inherited problems and the "newborn problems" but everything is still running on the same old lines as ever (cf. Weidner, 1991).

Empirical political studies have also made a major contribution to demystifying the ideology of all-encompassing state responsibility since they were able to demonstrate that the capacity, and indeed capability, of state controls to solve environmental problems in modern pluralistic industrial societies has been overtaxed. In view of this the leading scholar of the theory of environmental politics in Germany, M. Jänicke (1986), has claimed there is a structural failure on the part of the state in the field of environmental protection. In any case the reality of environmental policy in the Federal Republic of Germany has not always followed the standard bearers of a hierarchical, authoritarian interpretation of the role of the state. From the early days of systematic state environmental policy and over the years to an increasing degree there have been informal and also formal decision-making processes both on the level of programme conception and execution in which state organs were involved in negotiations and concluding agreements "simply" as representatives of particular interests. Their function often becomes that of state attorney, who legitimates the negotiated solutions of particular groups of societies, declaring them to be in the common interest (cf. Mayntz *et al.*, 1978).

Although negotiated solutions as such are hardly a question of dispute any more in the implementation of state obligations, and thus also in the field of environmental protection, they nevertheless have a bad reputation in the Federal Republic of Germany. From the pluralist and neo-corporatist point of view they are considered to be appropriate means of bringing about socially acceptable solutions to problems in complex, modern democracies but "environmentally-minded critics" point out an ecological blind spot in the pluralism model: interests which are shared by a large number of people (known as diffuse interests) are difficult to organise and therefore have little chance of success in a system organised on the basis of a pluralistic range of associations (cf. Beck, 1988).

In addition to this the negotiated solutions *to date* are also criticised for taking place under the false appearance of being under authoritarian state control. This, say critics, means that the huge potential of negotiation to make a contribution to flexible, appropriate solutions to problems which could reduce conflict is not used to full capacity, is even given a negative bias, thus causing unnecessary discrimination against this very promising method (cf. Scharpf, 1992).

Environmental activists and organisations as a rule view negotiating processes as something of a "fiddle", a secret pact between those parties to whom economic considerations are more important than what is ecologically required and possible.

The criticism they express is not, as they are often accused of, a general suspicion with no real specific basis. There are good reasons for their criticism of the interpretation and implementation of the co-operation principle which favours "polluter interests". A good number of negotiated solutions which, for example, were worked out in the form of what are known as sectorial agreements (i.e., concerning a particular industry) do little more than set down on paper what is already common practice, were simply not adhered to, or granted the companies causing environmental pollution generous deadlines for changing their policy (resulting in contaminated sites that had been given advance approval). The agreements with the automobile industry on reducing exhaust fumes and with the packaging and asbestos industries are prime examples of this (cf. Wicke, 1989, pp. 233 ff). Procedures which formally require public participation, especially those involving a large number of people and which are particularly rife with conflict, are generally perceived as "alibi procedures" in which the essentials have already been worked out between the government authorities and the representatives of the company concerned. Individuals and groups representing their own particular interests or the interests of the environment in general are on the whole excluded from the "preliminary negotiations" which normally take place before the formal procedure begins.

It is, therefore, no wonder that formal participation procedures, particularly those involving the general public in cases likely to cause a good deal of conflict (such as nuclear or conventional power stations, waste incineration plants, airports), are perceived as being a mere alibi, having nothing but a veneer of public participation and a preprogrammed outcome. It is thus completely logical that informal methods of protest are used and mobilisation of the law courts is seen as the most promising way of ensuring the public interest wins through. As far as implementation of legislation is concerned, both on a local and regional level, many people right across the political spectrum experienced in the period of controversial environmental policy in Germany what the government authority in question really thought of the co-operation principle (cf. Knoepfel and Weidner, 1985).

At central government level, where laws, regulations and the most important environmental standards are decided upon, it was primarily the environmental organisations which had bad experiences with participation procedures and also realised that the representatives of the state in the numerous standards committees and other committees could often not compete with the expertise available to the companies affected by environmental legislation. The inclusion of neutral scientific advisors was evidently not always able to shift the imbalance in favour of the environment. The independent committee of advisors on environmental questions – *Sachverständigenrat für Umweltfragen* – pointed this out: "It is not always possible to decide on a scientific basis what the optimal state of a particular environmental aspect is. It is far more important that society and the parties to the conflict are willing to seek a consensus on the quality of the environment to be aimed for in each case and the standards which therefore have to be imposed in democratic, participatory, and if necessary conflictual procedures . . . . The committee believes that, for the sake of acceptance by all concerned, both the procedure under which environmental standards are drawn up and the evaluation philosophy they are based on should be made more transparent." (Der Rat von Sachverständigen für Umweltfragen, 1987, p. 17).

It is the way people have experienced the various levels in the state hierarchy and the formal and informal procedures which has created a fundamental lack of trust in the professed desire of environmental policy-makers and administrators to act in the public interest; a mistrust which, in the meantime fed by developments in other fields of politics, has led to political parties and state organs suffering a general loss of credibility. It should be stressed in this context that the role of state representatives in negotiated solutions concerning the environment, which in my opinion has rightly been criticised, is not the result of the occupational hazard of wishing to manipulate at the cost of the interests of the environment but to a large degree the result of pressures caused by the very structures themselves to clothe what are actually very reasonable (and unavoidable) forms of behaviour in a disguise which is accepted in the conventional and normative philosophy towards the state and the legal system. In doing so they not only place themselves but also the procedures in an ambiguous light.

These institutional and normative starting conditions and the background of conflict and bad experience must be borne in mind when considering the chances of success of a new approach to environmental policy in the Federal Republic of Germany based on negotiation, a system which to judge by the current discussion of political theory and, as I have tried to show in the preceding chapters, on the basis of experience in other countries, is thought to have great merits.

6. Conclusion

Right across the board in the field of political theory there is a consensus amongst the most important representatives of the various fundamental positions; they all agree that the need for control of society has risen drastically, that the forms of control prevalent in the past are proving to be seriously inadequate, that non-hierarchical control methods which set out general guidelines should be used more and that finally the state should be more reserved in its desire to exercise control but should nevertheless not entirely abandon its overall responsibility for shaping society. An objective need for state control is identified mainly because it is not thought possible that the problems of the environment be solved in a purely "evolutionary" manner nor simply by unleashing market forces (cf. Offe, 1987). The lowest common denominator in the various estimations of the role and capacities of the state is the view that the state is in principle capable of exercising control, in the sense that it has the capacity for a conceptually orientated shaping of the social environment by its political institutions (Mayntz, 1987) on the condition that "yesterday's" forms of control are replaced by modern ones. It is above all the "soft" forms of control which are classified as modern: context control, control by procedure and by provision of information, consensus and dialogue-style arrangements.

In view of the increase in environmental conflicts it is particularly the consensus and dialogue style procedures for creating a common will and taking decisions which of all the control forms mentioned are enjoying the greatest renaissance. They represent ways of supplementing or substituting authoritarian and hierarchical decisions by conflict management techniques (particularly those which take into account minorities or groups in a weak legal position) which to a large extent aim at dialogue-style procedures and negotiated solutions and are usually guaranteed by agreements or articles. They are the expression of a modified interpretation of the state, since the state control organs in relationship to other social systems now occupy only the role of "*primus inter pares*" and grant the agents concerned the possibility of finding a common goal.

All the control forms mentioned, which are not pure "types" but to a greater or lesser extent overlap and make cross reference to each other, do not entirely replace conventional, hierarchical state intervention; on the contrary, they are to some extent dependent upon this intervention in that, for example, types of procedure, general framework conditions or the right to participate are themselves the subject of legislation. They do, however, leave the "fine tuning" to social sub-systems.

With regard to the discussion on environmental policy instruments (as well as on policy instruments in general) and the discussion amongst political scientists of state control instruments, it can be seen that *mediation procedures* have a great deal of common ground with the forms of control and instruments currently considered most suitable for dealing with problems and conflicts in advanced industrial societies. The conclusion can therefore be drawn that theoretically mediation procedures have great advantages over conventional environmental policy instruments. My experience so far with our mediation project, as well as the experience of mediator-assisted negotiations in other countries, leads me to believe that this is true if some constraints are taken into account: they are on the one hand not appropriate for settling disputes based on fundamental conflicts of values; on the other hand it should be noted that the capacities of non-governmental environmental organisations and citizens' pressure groups can very easily be strained by such time consuming procedures. It is not merely the time invested and financial aspects which cause the strain but also the high psychological stress placed upon the representatives of these groups when they try to commit their constituencies to agreements reached in the mediation process. These procedures in the arena of environmental politics will thus probably be confined to local and regional conflicts which have a direct effect on the immediate environment of those involved and are therefore of manageable dimensions.

I believe that mediation procedures, or more generally speaking alternative dispute regulation procedures, are potentially a valuable addition to the existing range of instruments for settling environmental problems. It would be exaggerated to expect them to be a fail-safe solution in the world of environmental policy. There will probably never be a single "path to heaven" of this kind, a super-policy instrument, in the highly complex field of environmental policy. This can be seen in all the practical experience of all countries to date, particularly those using the economic incentive instruments which were favoured for a time. In the (humble) realisation that not the search for *the* most appropriate super instrument but the pragmatic development (in a process of trial-and-error) of a wide range of different instruments – which permit a flexible reaction to the similarly wide range of problems – is the most promising approach, students of environmental policies when critically evaluating mediation procedures and theoretically considering different environmental policy instruments should not let their evaluation criteria and expectations become unrealistically high. They should similarly not be led astray by uncritical advocates of mediation procedures, particularly those who have a vested interest in these procedures and paint all too rosy pictures of the capabilities of these procedures.

Notes

1 I find it of academic interest to ask whether European countries are generally less innovative than the U.S.A. when it comes to "inventing" instruments for implementing environmental policy and, if so, why. On the other had, when it comes to implementation, some European countries seem to be more effective than the U.S.A.

2 "These new tools of environmental conflict management and environment decision-making are understood as supplemental in legislative, regulatory and judicial processes." (Crowfoot and Wondolleck, 1990, p. 13)

3 This was shown at the conference "Comparative Research on Environmental Administration and Policy-Making", Drobak (Norway, 11–14 June 1992; organised by Alf-Inge Jansen (University of Bergen).

4 The significance of informal procedures for dispute settlement in the field of environmental policy, especially waste management, is currently being examined by Japanese experts in cooperation with the WZB mediation project which includes a comparative study of six countries.

5 These experiences are specific to this procedure with all the features specific to this district, both in terms of the case in hand and the political situation. They are also specific to what is certainly an atypical situation, i.e. that this procedure is partly being financed by federal research funds (Federal Ministry for Research and Technology) and that it was initiated by a research institution. The following comments on the results have been prepared by my colleague, H.-J. Fietkau.

6 The evaluation was carried out by K. Pfingsten (WZB).

7 Cf. the institutions listed in the newsletter CONSENSUS, published by the Public Disputes Network, Cambridge, MA.

8 This task goes beyond what is normally thought of as "persuasion" in that the mediator is required to state very clearly the advantages and disadvantages of taking part or not taking part. Compare this to the following definition: "Persuasion takes place when A influences B to adopt a course of action without A's promising or threatening any reward or punishment. It may take the form of example, expectation, proposals, information, education, or propaganda."

9 According to C.W. Anderson, quoted by Howlett (1991, p. 2), for the theorist of policy instru-

ments, "politics is always a matter of making choices from the possibilities offered by a given historical situation and cultural context. From this vantage point, the institutions and procedures of the state to shape the course of economy and society become the equipment provided by a society to its leaders for the solution of public problems. They are the tools of the trade of statecraft."

10 An international conference in January 1993 in the WZB by Jänicke and Weidner analysed "successful examples of environmental policy". The results will be published in 1994.

11 Comments on this in H. Weidner: Basiselemente einer erfolgreichen Umweltpolitik. Ms. Berlin 1992. Also includes numerous cross-references to further reading.

References

Amy, D.J. (1983) The politics of environmental mediation, *Ecology Law Quarterly* **11**, 1–19.

Amy, D.J. (1987) *The Politics of Environmental Mediation*, New York.

Amy, D.J. (1990) Environmental dispute resolution: The promise and the pitfalls, in N.J. Vig and M.E. Kraft (eds), *Environmental Policy in the 1990s, Towards a New Agenda*, Washington D.C., pp. 211–234.

Bacow, L.S. and Wheeler M. (1984) *Environmental Dispute Resolution*, New York.

Beck, U. (1988) *Gegengifte. Die organisierte Unverantwortlichkeit*, Frankfurt a.M.

Bingham, G. (1986) *Resolving Environmental Disputes, A Decade of Experience*, Washington D.C.

Böhret, C. (1992) Zur Handlungsfähigkeit des funktionalen Staats der spätpluralistischen Industriegesellschaft, in B. Kohler-Koch (ed), *Staat und Demokratie in Europa*. 18. Wissenschaftlicher Kongress der DVPW, Opladen, pp. 116–129.

Cormick, G.W. (1980) The "theory" and practice of environmental mediation, *The Environmental Professional* **2**, 24–33.

Cormick, G.W. and Knaster, A. (1986) Mediation and scientific issues, *Environment* **28**(10), 6–30.

Crowfoot, J.E. and Wondolleck, J.M. (1990) *Environmental Disputes, Community Involvement in Conflict Resolution*, Washington D.C.

Dahl, R. and Lindblom, C. (1953) *Politics, Economics and Welfare*, New York.

De Soet, M.C. (1988) *From Competition to Collaboration*, Ministry of Transport and Public Works, The Hague, April 1988.

Dembart, L. and Kwartler, R. (1980) The Snoqualmie River conflict: Bringing mediation into environmental disputes, in R.B. Goldman (ed) *Round-Table Justice: Case Studies and Conflict Resolution*, Reports to the Ford Foundation, Boulder, Col. 39–58.

Der Rat von Sachverständigen für Umweltfragen, Umweltgutachten 1987, Stuttgart-Mainz 1987.

Deutscher Bundestag, Referat Öffentlichkeitsarbeit (ed) (1990) *Schutz der Erde: Eine Bestandsaufnahme mit Vorschlägen zu einer neuen Energiepolitik*, 2 volumes, Reihe Zur Sache 19/90, Bonn.

Dierkes, M. and Fietkau, H.-J. (1988) *Umweltbewußtsein – Umweltverhalten*, Report for the Rat von Sachverständigen für Umweltfragen, Stuttgart.

Ellwein, T. "Staatlichkeit" im Wandel, Paper given during the plenary session of the 18th Congress of the Deutsche Vereinigung für Politikwissenschaft on 8.10.1991 in Hannover, Manuscript.

Eubel, P. *et al.* (1979) *Das Japanische Rechtssystem*, Frankfurt a.M.

Fietkau, H.-J. and Weidner, H. (1992) Mediationsverfahren in der Umweltpolitik, *Erfahrungen in der Bundesrepublik Deutschland, in Aus Politik und Zeitgeschichte* **B 39–40/92**, 24–34.

Gross, B.M. (1972) Political process, *International Encyclopedia of the Social Sciences* **11/12**, New York, London (reprint edition), 265–273 (here: 269).

Hansmeyer, K.-H. and Schneider, H.K. (1990) *Umweltpolitik. Ihre Fortentwicklung unter marktsteuernden Aspekten*, Göttingen.

Hoffmann-Riem, W. and Schmidt-Aßmann, E. (eds) (1990) *Konfliktbewältigung durch Verhandlungen – Informelle und mittlerunterstützte Verhandlungen in Verwaltungsverfahren*, 2 volumes, Baden-Baden.

Holznagel, B. (1990) *Konfliktlösung durch Verhandlungen*, Baden-Baden.

Hoppe, W. and Beckmann, M. (1990) Planfeststellung und Plangenehmigung im Abfallrecht, *Vorschläge zur Erleichterung des abfallrechtlichen Zulassungsverfarhrens, Reihe Umweltbundesamt*, Texte 27/90, Berlin.

Howlett, M. (1991) Policy instruments, policy styles and policy implementation: National approaches to theories of instrument choice, *Policies Studies Journal* **19**(2), 1–21.

Iklé, F.C. (1972) Negotiation, *International Encyclopedia of the Social Sciences* **11/12**, New York, London (reprint edition), 117–120.

Jänicke, M. (1986) *Staatsversagen, Die Ohnmacht der Politik in der Industriegesellschaft*, München and Zurich.

Jann, W. (1981) Kategorien der policy-forschung, *Speyerer Arbeitshefte* **37**, Speyer.

Kaufmann, F.X. and Rosewitz, B. (1983) Typisierung und Klassifikation politischer Maßnahmen, in R. Mayntz (ed) *Implementation Politischer Programme II*, Opladen, pp. 25–49.

Knoepfel, P. and Rey, M. (1990) *Konfliktminderung durch Verhandlung: Das Beispiel des Verfahrens zur Suche eines Standortes für eine Sondermülldeponie in der Suisse Romande*, in W. Hoffmann-Riem and E. Schmidt-Aßmann, pp. 257–286.

Knoepfel, P. and Weidner, H. (1985) *Luftreinhaltepolitik im internationalen Vergleich,* vol. 2, Bundesrepublik Deutschland, Berlin.

MacDonnell, L.J. (1988) Natural resources dispute resolution: An overview, *Natural Resources Journal* **28**(1), 5–10.

Mayntz, R. *et al.* (1978) *Vollzugsprobleme der Umweltpolitik*, Stuttgart.

Mayntz, R. (ed) (1980) *Implementation Politischer Programme*, Empirische Forschungsberichte, Königstein/Taunus.

Mayntz, R. *Politische Steuerung und gesellschaftliche Steuerungsprobleme – Anmerkungen zu einem theoretischen Paradigma*, in T. Ellwein *et al.*, pp. 89–110.

Offe, C. (1987) Die Staatstheorie auf der Suche nach ihrem Gegenstand. Beobachtungen zur aktuellen Diskussion, in E. Ellwein *et al.* (ed) *Jahrbuch zur Staats- und Verwaltungswissenschaft,* Vol. 1, Baden-Baden, pp. 309–320.

Pfingsten, K. (1992) Miteinander reden . . . ! Mediationsverfahren zur Bewältigung von Altlastkonflikten, in *Neue Zürcher Zeitung*, 26.8.1992.

Pfingsten, K. and Fietkau, H.-J. (1992) Mediationsverfahren: Leitgedanken und methodische Erfassungsmöglichkeiten, Darstellung der empirischen Erhebungsverfahren im Forschungsprojekt "Mediationsverfahren im Umweltschutz", FS II 92–305, WZB.

Ritter, E.H. (1987) Umweltpolitik und Rechtsentwicklung, in *Neue Zeitschrift für Verwaltungsrecht* **6**(11), pp. 929–938.

Roth, R. and Rucht, D. (eds) (1991) *Neue soziale Bewegungen in der Bundesrepublik Deutschland*, 2nd revised and expanded edition, Bonn.

Sagami, Y. (1989) The environmental pollution dispute settlement system, in S. Tsuru and H. Weidner (eds) *Environmental Policy in Japan*, Berlin, pp. 196–206.

Scharpf, F.W. (1992) Die Handlungsfähigkeit des Staates am Ende des zwanzigsten Jahrhunderts, in B. Kohler-Koch (ed), *Staat und Demokratie in Europa*. 18, Wissenschaftlicher Kongress der DVPW, Opladen, pp. 93–115.

Siebert, H., Interview in: Der Spiegel 33, 14 August 1989, pp. 33–50.

Striegnitz, M. Mediation. Lösung von Umweltkonflikten durch Vermittlung – Praxisbericht zur Anwendung in der Kontroverse um die Sonderabfalldeponie Münchehagen, *Zeitschrift für angewandte Umweltforschung* **3**(1), 51–62.

Susskind, L. and Cruikshank (1987) *Breaking the Impasse: Consensual Approaches to Resolving Public Disputes*, New York.

Susskind, L. and Madigan D. (1984) New approaches to resolving disputes in the public sector, *The Justice System Journal* **9**, 179–187.

Susskind, L., Bacow, L. and Wheeler, M. (eds) (1983) *Resolving Environmental Disputes*, Cambridge, Mass.

Tsuru, S. and Weidner, H. (eds) (1989) *Environmental Policy in Japan*, Berlin.

Umweltbundesamt (ed) (1992) *Daten zur Umwelt 1990/91*, Berlin.

Upham, F.K. (1987) *Law and Social Change in Postwar Japan*, Cambridge.

Van den Dale, W. (1992) The research program of the section "Normbildung und Umwelt", FS II 92–301, WZB Berlin.

Weidner, H. (1991) Umweltpolitik – Auf altem Weg zu einer internationalen Spitzenstellung, in W. Süß (ed) *Die Bundesrepublik in den Achtziger Jahren*. Innenpolitik, Politische Kultur, Außenpolitik, Opladen, pp. 137–152.

Wicke, L. (1989) *Umweltökonomie, Eine Praxisorientierte Einführung*, 2nd edition, Munich, p. 233 ff.

Willke, H. (1991) Regieren als kunst systemischer intervention, in H.-H. Hartwich and G. Wewer (eds) *Regieren in der Bundesrepublik III*. Systemsteuerung und "Staatskunst", Opladen, pp. 35–51.

9. New Institutional Arrangements for the Next Generation of Environmental Policy Instruments: Intra- and Interpolicy-Cooperation*

PETER KNOEPFEL
IDHEAP, Université de Lausanne, Lausanne, Switzerland

1. Introduction

It is hardly surprising that the European Science Foundation chose environmental policy instruments as the topic of one if its task forces on 'environmental policy'. All contributions to this volume testify that environmental policy lacks the confidence that it can achieve its goals with the classical environmental policy instruments. After twenty years of hard work of implementation many policymakers arrived at the conclusion that its instruments have to be supplemented by new, more efficient tools. It seems that after a swift booming phase in the late 1970s and 1980s environmental policy, in the light of its current overarching political importance, has already for some time considerably transgressed the simple frame of ordinary policing in the field of health or industrial supervision. More recently, in the search for new instruments, theory and praxis environmental policy has borrowed increasingly from fiscal and tax policy.[1] Moreover, the slowly adopted practice of environmental impact assessments had led to first attempts to build a bridge with industrial, land use and infrastructure policy.[2] The EIS requires the application of environmental law in the planning and realization of new projects of all governmental policies of environmental relevance. The postulate has always been included in all the relevant laws and at least since the Single European Act of Luxembourg (1987) it figured in the pertinent body of primary law of the European Community (Art. 130 R, Para. 2). In the Maastricht Treaties it received an additional normative reinforcement.[3]

Everyone who deals with such EIS procedures is well aware of the enormous resistance against these high principles in practice. Neither the EC guideline no. 85/337/EEC of June 27, 1985, nor other known national environmental legislation within and outside of the European Community envisage legally binding, unambiguous regulations stating that facilities cannot be built if the EIS is negative. Nevertheless, the European-wide introduction of the EIS produced the important

Bruno Dente (ed), Environmental Policy in Search of New Instruments, 197–233.

insight that the most substantial innovation of this instrument is the creation of new administrative structures inside and outside of the traditional environmental bureaucracies, which facilitate coordination.

This chapter will however not focus on the EIS which is still an instrument of classical environmental policy, especially if it is just introduced into already existing permit procedures. Since it focuses on specific projects and operates based on a relatively narrow, traditional understanding of environment, this procedure could be accommodated, at least in the initial phase, within the existing administrative structures. It hardly produced systematic cooperation between environmental policies and the core-actors of infrastructural supply policies on a strategic level. At the centre of this article are the strategies leading to a lasting, systematic and innovative opening of the non-environmental policies towards ecological policies in the comprehensive sense. Such an extension would among other things improve the access of ecological policy to the instruments of implementation within these non-environmental policies. The spectrum of ecological policy instruments would therefore be enlarged not only in the direction of specific economic instruments of fiscal policy, but in the direction of other regulative, incentive or persuasive instruments of potentially all realms of governmental public policies.

The realization of this ambitious goal presupposes the more or less far-reaching inter-linkage of environmental policy with these policies. Such rearrangements are the main topic of this article. In a first chapter, the potential scope of this new set of instruments is explored on the basis of three examples (agriculture, transport and land use policies). Some considerations are made regarding the price the environmental policy would have to pay to the non-environmental policies for such a borrowing of instruments (Chapter 2). In a second chapter, the empirical case of Switzerland is briefly discussed, which will serve as an illustration for developing scenarios for restructuring the environmental administration (Chapter 3). The centre piece of the argument is laid down in Chapter 4, in which scenarios are presented for reinforcing cooperative structures within an environmental policy, which is expanded in regard to its sphere of application (intra-policy cooperation), and for the extending of instruments respectively, which is based on various concepts of structurally assured cooperative relationships between environmental and non-environmental policies (inter-policy cooperation). This chapter is entirely drawn from an expertise Willi Zimmermann and I worked out for the attention of the Swiss Parliamentary Control Commission on the Federal Office of the Environment, Forests and Landscapes (FOEFL) in 1991.[4] In the final Chapter 5, these scenarios are briefly referred to with regard to their political and instrumental implications.

2. The New Potential Spectrum of Instruments and its Price

Quite naturally, most of the instruments of the non-environmental policies, which are of interest to ecological policy consist of (direct) planning and execution in their implementation phase. At the level of program formulation *per se*, some binding obligations for cooperation of a material or a procedural nature can be envisaged. In most cases, these obligations remain however irrelevant for the potential changes of behavior of the interested target groups as long as there are no (judicial) sanctions in case of non-compliance (for instance an invalidation). It has to be said, however, that for the extension of the access of the environmental policy to the federal or (above all) regional (= cantonal) political-administrative institutions which apply these instruments some legislative acts are also required. They pertain to the restructuring of existing administrative arrangements at the organizational and procedural level. The most current case is the creation of new regional implementation agencies. They are the key precondition for making accessible the vast field of existing policy instruments to the political-administrative actors of environmental policy with sufficient institutional guarantees. These complementary strategies are the main focus of this article.

Before discussing these strategies, a brief description of the issue areas and policy instruments is necessary. Of environmental relevance are all those non-environmental policies, which are objectively in a position to affect substantially the environmental behavior, defined in space and time, of the target groups of environmental policy. These include, amongst others, all those policies of the federal state or the regions, which can only be realized through the construction of facilities. In the case of Switzerland, this includes the general policy of federal buildings, defense policy (barracks, shooting and training grounds, airports), road construction policy (national highways and main streets), the general policy of public transportation (federal railways, buses, shipping lines and air traffic), telecommunication policy (construction of transmitters), energy policy (construction of power plants, grids, pipelines), forestry policy (preservation of Swiss forests), environmental policy in the narrow sense (construction of sewage and waste disposal plants), housing policy, regional economic policies (development of mountainous areas), tourism policy and, last but not least, the large field of land use policy.[5] In addition, there are policies, which are of indirect relevance for the environment, such as agricultural, research or education policy.

Given this almost unlimited wealth of environmentally relevant, non-environmental policies and the related, individually applied instruments, every strategy of ecologizing non-environmental policies has to set priorities. Concentration on a few policy areas, which will probably vary over time, is mandatory in the interest of practical applicability and political feasibility. Most likely, agricultural, transport and land use policy today form part of this core area in Switzerland and other Central European States:

- The tools of agricultural policy comprise regulative instruments such as import restrictions, limitations of feed stuff, production restrictions through the licensing of agricultural building, or the limitation of livestocks through prohibition, as well as the immensely wide field of agricultural product subsidies or product-unrelated direct payments tied to corresponding conditions of extensive production, and the various types of agricultural consulting (persuasive instruments).[6] Compared with other policy areas, agricultural policy has a wealth of instruments deriving from years of experience, which also reflects the manyfold corporatist ties with various groups of agricultural policy (food stuffs industry, producers of agricultural subsidiary products, large, small and mountain farmers, various branches of the food industry). Accordingly, the implantation of concrete ecological elements through an explicit integration of the application of these instruments in the realm of environmental policy would facilitate a multifarious, faceted and differentiated implementation of environment along the entire spectrum of food production starting from the raw materials, to the farms and to the product sector.[7]
- In transport policy, in most countries today, the regulative instruments prevail both with regard to the construction as well as the operation of the transportation network. With regard to the construction of these works, plans and project approval procedures dominate in different phases. In addition, there are of course various budgetary decisions allocating the corresponding investments. With regard to the operation of these systems, the manyfold instruments of regulative traffic legislation (for instance, signalization of speed limits, (partial) driving restrictions, neighborhood streets etc., as well as their control)[8] and also measures of traffic limitation through construction authorization procedures (road blocks, bars, etc.) are of increasing interest to environmental policy: More recently, road traffic policy increasingly uses incentive instruments, which promise fiscal rewards for low exhaust vehicles or reduced driving (and the corresponding transfer to public transportation). Other essential instruments include guidelines for the equipment of vehicles of private and public transportation.

The air hygiene action plans designed by the Swiss cantons in order to reduce air pollution by road traffic provide an impressive illustration of the variety of instruments of regional transport policy which are of interest to environmental policy, and their consequent operationalization by environmental policy. First analyses of the genesis of these plans illustrate that such a building of bridges between environmental and transport policy is only successful if the corresponding organizational and procedural rearrangements at the level of the responsible administrations an be institutionalized permanently.[9]

- Surprisingly, in Switzerland and some other European countries, the instruments of land use policy have only been used occasionally for environmental policy. In the case of Switzerland, reasons include the different organization of federal competencies in these two policy areas, the different professional standards, the delegation of these activities to two different federal agencies, and last but not least the quite different nature of the instruments themselves. Nevertheless, the land use policy instruments (such as zoning) would be of substantial interest to environmental policy, since they ultimately decide the nature, the spatial distribution and the density of potential sources of emission. Key works such as 'condensed construction', 'space saving road planning', 'protection through allocating of compensatory areas close to nature', 'biotops' and 'nature protection objects', or 'definition of livestocks ceilings conducive to the environment in agriculture' are clear illustrations. In Switzerland, there only have been occasional forms of cooperation between these two policy areas so far. Examples include noise protection (identification of noise sensitive areas), environmental impact assessments (definition of the spatial perimeter within which the impacts of new projects have to be examined), action plans for air protection, or (newly required) environmental impact statements for communal land use plans. But environmental policy has not yet advanced into the core areas of zoning policy (housing, transport, and industrial zoning policy). A much more comprehensive cooperation and a corresponding utilization of instruments by environmental policy would be possible.[10]

It is obvious that the non-environmental policies concerned and the responsible authorities, and also the interest groups standing behind them, will not easily put 'their' instruments at the disposal of environmental policy. They would have to concede much more to the environmental policy than they do today in the frame of the ultimately non-binding environmental impact assessments, where they still retain the final decision on the realization or non-realization of a public work. The extension of their instruments to the environmental side would at least imply that environmental policy would not gain a right of initiative, but an (absolute?) veto position. For this purpose, environmental policy would have to pay a price, since the policies concerned would be 'slowed down' or even put into question not only at the fringes, but occasionally at the very center of activities.[11]

Because instruments are not only unilateral acts of authority, and because there are often extensive negotiations with the target groups before their application, the non-environmental authorities have also to partially give up their monopoly-like control over the various contacts to these target groups in favor of environmental policy.

Stiff resistance by the non-environmental policies can be expected. It can only be broken, if environmental policy in turn can offer advantages to the policies concerned. Such advantages exist indeed: They can be summarized under the title 'certificate of ecological compatibility'. Such a certificate would provide to the non-environmental authorities and their projects ecological legitimacy, safety and a favorable 'climate of allocation'. Today, there will hardly be agreement over the political opportunity of issuing such certificates of compatibility. This problem will be addressed again in the context of the presentation of the scenarios in Chapter 4. At this point, it only has to be said that with this offer in return environmental policy assumes a high, considering the lack of knowledge in some areas, even very high, liability. Put into question later on, it would not only incur high public legal costs ('protection of trust'), but would also undermine the pact of 'do ut des' between environmental and non-environmental policies described above. The danger has to be considered as relatively high that given such premises the utilization of the instruments of these other policies would be refused to environmental policy from the beginning.

3. The Empirical Background: Analyses and Recommendations Regarding the Restructuring of the Swiss Federal Office of the Environment, Forests and Landscapes (BUWAL/FOEFL)

The above mentioned review of the Federal Office of the Environment, Forest and Landscapes (BUWAL/FOEFL) conducted during 1991 on behalf of the Control Commission of the Swiss National Council constitutes the empirical background of the scenarios presented in this article.

In interpreting the findings the authors followed inductively and deductively the following eleven guidelines in the sense of grounded theories:

"1. The structures of the agency should altogether be simplified and the spectrum of tasks (objectives, functions) be reduced. It should have a basic structure which corresponds better to the requirements of an integrated ecological policy through internal linkages than it does today.
2. Improved internal linkages exist, if the agency is built according to the basic principle of a matrix organization. This form of organization allows to link vertically the sectorial departments with the various environmental media (water, air, noise, nature, landscape, soil) through so called cross-sectorial department. The cross-sectorial departments regulate issues which – comparable to ecological circulations [*Stoffkreislauf*] – circulate through several environmental media and thus pertain to the competencies of several sectorial departments.

3. According to a clear distribution of competencies, the technical responsibilities should be assigned to the departments operating close to the problem. The level of main departments has to be abolished, since these units produce no visible efforts of coordination and thus form a redundant intermediate-level at least in the areas under closer empirical scrutiny. These departments should be integrated into the two groups of the (already existing) sectorial departments and the cross-sectorial departments yet to be created.
4. The sectorial departments have to be structured into emission-oriented sections (following the example of the Clean Air Department). This means that for each major group of polluters (= [*Emittenten*] a section exists [Ansprechpartner]. The competencies of these sections for the conduct of selected affairs have to be defined clearly and to be adjusted periodically to changing conditions.
5. In the foreseeable future, the task of the agency should be strictly limited to the performance of environmental *protection* functions, in order that the agency can encounter the utilization functions administrated by other agencies with the same consequent strategy. A simultaneous representation of protection and utilization functions in one and the same agency creates unnecessary conflicts of interest which are detrimental to the resolute implementation of the agency's legal mandate.

   Such a solution is likely to be criticized on the grounds that potential conflicts between parts of the federal administration cannot be dealt with inside the agency as in other cases. The Ministry, essentially the already overburdened Head of the Ministry (the Federal Counsellor), would have to decide once more. The integration of different points of view should be achieved at the level of the agencies; it should be part of the quality of a agency head that he can resolve such tensions internally.

   A reply to this frequently aired reservation is that given the consequent implementation of the legal mandate of protection, for which in most cases – namely the definition of norms – scientific and not political arguments are decisive, there is hardly room for compromise at the operational level. Where they are necessary, they should be taken at the strategic-political level of the corresponding Ministry and clearly earmarked as political decisions. Where required by the political implications of such conflicts, regulations should be taken in the framework of environmental delegations by the Federal Council (environment/transportation; environment/agriculture/forest; environment/energy, etc.). This concentration on the essential mandate of protection corresponds to the existing basic structures of the federal environmental administration. It is mandatory as long as the protection functions cannot be reliably performed by the other federal agencies (for the future, see the scenarios in Group B in Part III of this report).

6. Strategic leadership of the agency has to be reinforced in a way which results in a reduction of the duties of the Head of the Ministry (Federal Counsellor). For this purpose, the internal leadership capacities of the Direction (which has been described by all interviewees as too large and too preoccupied with details) should be improved and more operational decision be delegated to the departments.
7. The FOEFL currently employs approximately 320 persons. Such a large agency needs an independent strategic-operative management located outside of the specialized departments which can rely on the functioning structuring of the competent departments. A directive and reductive management is required. Leaders should set priorities and assure their implementation through protective measures against competing demands. For this purpose, procedural rules and guidelines have to be developed.
8. Multifunctional service branches or staff should be avoided. Staffs should fulfill either person-, issue- or performance-oriented tasks. Given the size of the agency, a mixture of these functions (in one and the same branch) leads to lack of transparency, inefficiency and frustration on the part of the employees working in the line.
9. The extremely valuable resource of a highly qualified and motivated staff should be used and not abused according to the ecological principle of sustainability.
10. In the cross-sectorial departments in particular, more generalists and social scientists trained in ecology (especially economists, political and administration scientists and psychologists) should be employed. They should not be recruited for particular issues or functions. Rather there has to be more rotation in order to avoid structures being developed according to personality attributes of the existing personnel and not according to issues.
11. Wherever competition allows, where qualified for the performance of the agency's tasks, and available on the open market, such services should be hired from the outside according to market conditions."[12]

In the context of this report, guideline no. 5 in particular requires a brief commentary: Proposing a strict concentration of the agency's tasks on the realm of environmental protection and the related out-placing of the current resource utilization functions in the realm of the forests, the report clearly puts emphasis on improving the coordination of the various environmental policies (intra-policy cooperation). Accordingly, in a first step, all environmental protection policies should be concentrated in one agency and implemented according to a unified strategic philosophy of intervention; concerns of resource economy, which potentially weaken this philosophy, should be performed by other agencies.

The report argues that no exception from this principle should be made in the case of the forest which, in regard to protection and utilization, are administered by the FOEFL. While the utilization of the forests has also a 'social function', this is the case as well in other policy areas such as energy, agriculture or water economy, which also have to perform a social (respectively an ecological) function; 'without policy conceding [them] in a special statute an ecologically legitimized self-regulation . . .'.[13] As the report continues thee could be arguments against this proposal that 'the current forest model (against the opinion of the expert) is not a relic of times past, but rather a nucleus for a future new pattern of ecological administration. The point is to integrate the utilization and protection of resources in one and the same administrative structure. In fact this is an interesting scenario for the future . . . . For the near future, however, this model is hardly applicable, since there are no obvious reasons to transfer other, ecologically as important or even more important, utilizations of resources such as agriculture or water economy to the FOEFL.'[14]

The aforementioned report continues: 'For the administrative power of the environmental policy at the federal level demanded by the Control Commission on her behalf, the uniformity of the philosophy of strategic intervention is more important than the special promotion of the ecologization of one particular area which when compared with others (water, agriculture) is ecologically no more relevant. Either the entire agency switches to a strategy of administering ecological resource utilization in general at the level of the philosophy of strategic intervention [. . .], or it uniformly concentrates on the protection function. The coexistence of both philosophies in one and the same agency is detrimental to the development of a uniform strategy and a shared 'corporate identity'.'[15]

It is no surprise that this proposal has not been spared criticism in the public debate. The Control Commission only partly followed this recommendation.[16] At the center are different evaluations of the optimal time needed to switch from a protection- to an utilization-oriented environmental policy. The authors of the report, who themselves developed the transitional scenarios presented in Chapter 4 of this article in the last part of their report, were and are of the opinion that the optimal time for a systematic transition from protection to utilization in Switzerland's environmental policy has not yet come. Over the short to medium term (five to eight years), inter-policy cooperation has to be achieved on the basis of functionally clearly divided responsibilities. Too fast a transition to models which integrate protection and utilization functions in one and the same administrative institution, will result in a preponderance of the utilization and thus in a weakening of the protection functions, as is the case in the view of the authors in the realm of forest administration. The advocates of a quicker pace in turn emphasize the positive ecologization effects of integrating issues of forest economy into the FOEFL, which they claim to have observed since the integration of the forest into this agency in 1989.

The proposed reorganisation of the FOEFL concentrates as indicated on short-term improvements in the field of intra-policy cooperation. These proposals are summarized in Figure 1.

These proposals focus on the creation of cross-sectorial departments which will be linked to the concerning sectorial departments according to a matrix organization. The Direction will be streamlined, the intermediate level of the main departments abolished and their operational competencies delegated down to the level of the departments. The table also shows that the utilization-oriented affairs of the forest administration will be transferred to the Federal Office of Agriculture. The protection-oriented affairs of traffic regulation in turn should be newly subordinated to the FOEFL instead of the Federal Justice and Police Department. Finally, the report proposes to place the National Hydrology and Geology of Switzerland outside of the FOEFL and – in accordance with the corresponding concepts in the European Community and other European countries – to transfer it to a newly created Federal Agency for Observation of the Environment. The parliamentary Control Commission has meanwhile adopted most of these proposals (except the proposi-

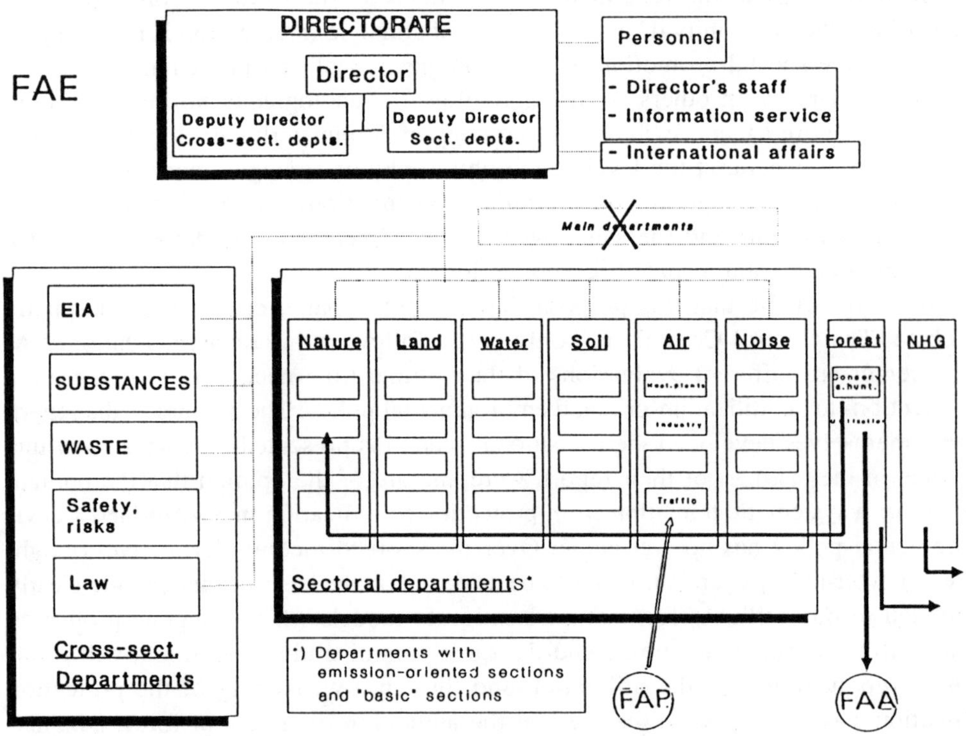

Fig. 1

tion to transfer the forest administration).[17] The proposed structure is presented in Figure 1.

The report includes the following definitions of the sectorial departments, the cross-sectorial departments and the proposed matrix organization which might be interesting for concepts of intro-policy cooperation in reform debates abroad:

"– The six sectorial departments observe and regulate the most important objects of protection according to the classical division of the most important environmental media as they already exist within the FOEFL (nature, landscape, water, soil, air and noise). Their main task consists of protective measure in favor of these media, which they best organize according to the various producing groups (and not according to more specific targets of protection such as ground or surface waters etc.). A second task, which is becoming more important given the discovery of numerous growing interdependencies between the various environmental media, consists of developing concepts of protection of specific ecosystems across the media. This should be achieved through flexible arrangements of regulation among several sectorial departments.

  Also increasingly important are cooperation arrangements with particularly relevant 'non-environmental' federal policies, which represent utilization interests outside of the agency (e.g. clean air and energy; soil protection and agriculture etc.). Based on existing material, cooperation duties, within the matrix structure (which has yet to be created) necessitate new cooperation agreements which have to be concluded and maintained at the junctions between the cross-sectorial departments concerned and the sections respective departments. The Basic Sections, which should exist in every department (data collection, data analysis etc. in close cooperation with the proposed 'Federal Agency for the Survey of the Environment'), should basically be designed according to existing examples (the Clean Air Department, for instance).

– The cross-sectorial departments analyze and regulate those environmental burdens which, through known substances or flows of substances or as a consequence of large projects, concern on a regular basis several environmental media in a way more or less calculable today (waste, substance, EIT). The effective regulation of these environmental burdens requires first a highly specialized, substance- or project-specific technical knowledge, which cannot be accumulated on the media-side. Second, on the side of the media, this regulation requires permanent opposites in the sectorial departments (so called 'junctions' in the matrix organization).

  These cross-sectorial departments can be structured into sections either according to specific substance- or issue-groups, or (following a more spatial view of regionally specific flows of substances) according to national regions. In both cases, the creation of basic sections is recommended, and these have to deal among others with economic aspects (e.g. international harmonization). The Law Department has a special cross-sectorial function, which will increasingly have to focus on the terminological etc. unity of the development of environmental law given the growing wealth and diversification of partial sectorial regulation by the sectorial and cross-sectorial departments.

– The matrix organization facilitates the coexistence of two departmental groups, whose competencies have been strengthened compared to the present situation, with a minimum of hierarchical guidance and bureaucracy through a concept of pre-structured, horizontal cross-linkages, which in turn are organized under the guidance of a department (and are thus not completely 'random'). Experience shows that a matrix structure cannot be 'designed exclusively on the drawing-board'. In particular, the idea of flexibility facilitating fast cross-linkages should not be turned into the opposite of all-round blockade through perfectionist bureaucratization. Rigid patterns of organization should be avoided and flexible, task-oriented arrangements between individual departments should be promoted outside of the (continuous) cross-sectorial departments. These 'task forces' need however a clear assignment and they should be approved by the Direction (without such an approval, there is the – well-known – danger of 'flying bureaucratic carpets', which constitute, legitimize etc. themselves).

Such a matrix organization is an important organizational instrument for implementing an environmental policy, which is increasingly oriented towards the ecological overall context (and not toward individual laws and directives). It is also likely to form an important precondition for developing a future generation of more planning-oriented executive instruments on behalf of the cantons."[18]

In the following Chapter 4, we present the two groups of scenarios A (intra-policy coordination) and B (inter-policy cooperation) as they appear in the FOEFL report. In this chapter, we intentionally reproduce the text of the experts in a direct translation without any further comments because this (publicly less debated) piece of our report seems to be of interest to most European countries aiming at restructuring their environmental agencies. They are formulated in a neutral way leaving all political considerations up to the political decision makers. For the purpose of this article, however, we allowed ourselves to add in Chapter 5 some personal comments on their political implications and a brief analysis of their potential contribution to an enlargement of environmental policy instruments. In the final chapter, they will be briefly commented on in respect to the problem of environmental policy instruments, which are of a particular interest in this article (Chapter 5).

On behalf of the politicians concerned, the report starts with the following observation: "Current political debates on future environmental policy, as a rule, focus on either the expansion of its protective functions or a general expansion of its scope, namely, a reorientation from traditional environmental protection to environmental uses. There is, for example, a widely shared opinion that the objects to be protected by environmental policy should be defined more precisely, studied more thoroughly and better taken care of with the help of appropriate measures based on the concept of integrated policies. Furthermore, claims are made for an expanded definition of the term 'risk' in order to protect the environment from sources of risks that hitherto have received too little attention. A probably equally important category of new political concepts considers the restriction of environmental policy to protective functions per se as inappropriate, because it ultimately hinders environmental policy from taking a direct influence on the cause of a particular environmental use. Environmentally problematic policy areas, by comparison, have more pull due to their long-standing tradition, their good supply of funds and personnel, and their strong support from powerful interest groups. Therefore, it is argued, implementing environmental policy that is restricted to protection is bound to be a Sysiphean task. Consequently, environmental policy should be given better access to non-environmental policy areas wherever these are responsible for the rise of environmental damage.

The expansion of protective functions as well as the integration of environmental uses into the realm of environmental policy do not only require numerous modifications with regard to the contents of legislative programs, etc. but also the reorganization of the administrative structure of environmental policy. Since this expertise concentrates on the latter, we shall present in the following a set of scenarios of administrative structures that are based on the various political options outlined above. Needless to say that the implementation of a new environmental-political concept cannot start with reorganizing the administration. First, a political decision has to be made on one or the other possible political objective. In the following eight scenarios these objectives can be presented only in rough outlines because a detailed justification and analysis would not be possible within the frame of this expertise."[19]

4. Scenarios for Future Rearrangements

4.1. *Two Scenarios for Strengthening Intra-Policy Cooperation*

The first two scenarios deal with the restructuring of the present Federal Agency of Environment, Forestry and Agriculture (BUWAL) into a new Federal Environment Agency (BFU). Scenario A1 illustrates those changes within the agency to be

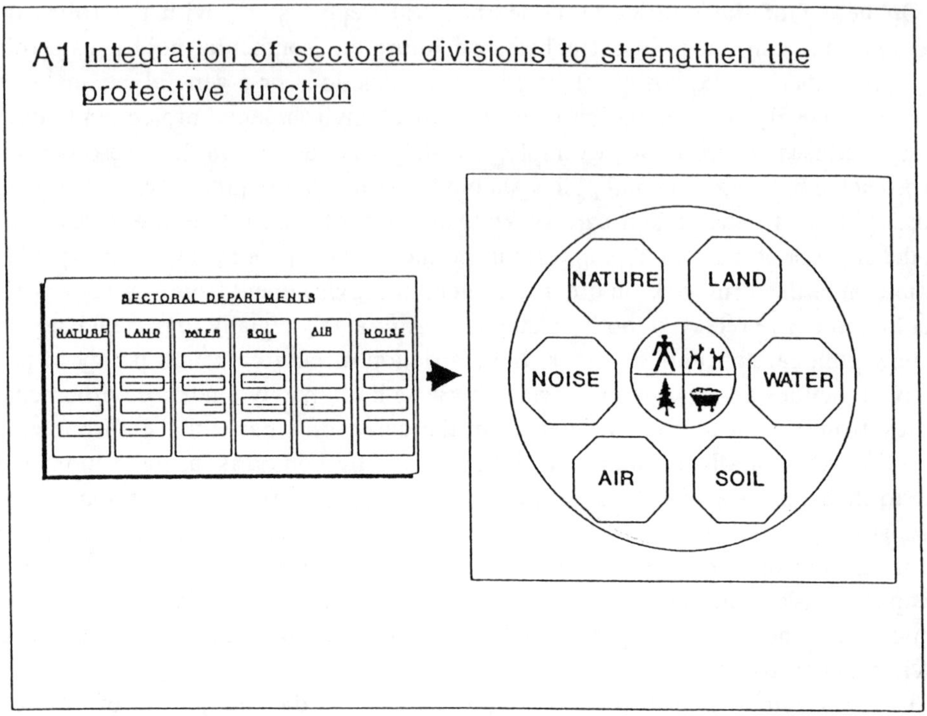

Scenario A1: Integration of a new sectoral division to strengthen the protective function.

recommended, if an expansion of the protective functions of environmental policy is aimed for.

The present environment agency (BUWAL) consists of sectoral divisions that are each specialized in one environmental medium (air, water, soil, etc.). Scenario A1 envisages a better interconnection of these divisions by grouping them around a common core division, thereby bringing them close to the actual objects of protection instead of the environmental media that are now being protected on behalf of the objects. As laid down in the federal Environmental Protection Act, objects of protection are human beings, animals, plants, and materials. By now, each sectoral division, in performing its tasks of protection, has of course adjusted itself to these objects. Within the administrative structure, however, the objects themselves and their needs are represented only implicitly, and there is as yet no central division that could, so to speak, act as an agent and plead the cause of these objects in their capacity as legal entities. Scenario A1 has the additional advantage that synergistic effects (e.g., air and noise; water and biotopes, etc.) could be countered more effectively because in their balance sheets of suffering the objects of protection, as a rule, do not differentiate among environmental media.

The opponents of such administrative reorganization argue that the interests of protected groups are represented by other large federal agencies. As for human health, reference is made to the Federal Public Health Agency; as for animals, to the Federal Veterinary Agency (located in the Federal Department of Economics; as for materials, to the Federal Institute of Material Testing and Experimentation (EMPA) or to the Federal Agency of Culture (regarding historical monuments). However, a closer look at these agencies reveals that environmental risks and dangers do not play a key role either in the policy of these agencies or in the legislation they administrate. The Federal Public Health Agency, for example, adopted the topic 'environment and health' (within its section 'medical epidemiology'?) only about two years ago.

Scenario A2 deals with the level of transversal (cross-medial) divisions within the environment agency and is based on an expanded definition of the term 'risk'. In this sense, it would, for instance, become necessary to consider the integration of radiation protection – today largely performed on the basis of self-control – according to, e.g., the German model (Responsibility for radiation protection currently lies with the Federal Energy Agency and the General Secretary of the Department of the Interior = GSEDI).[20]

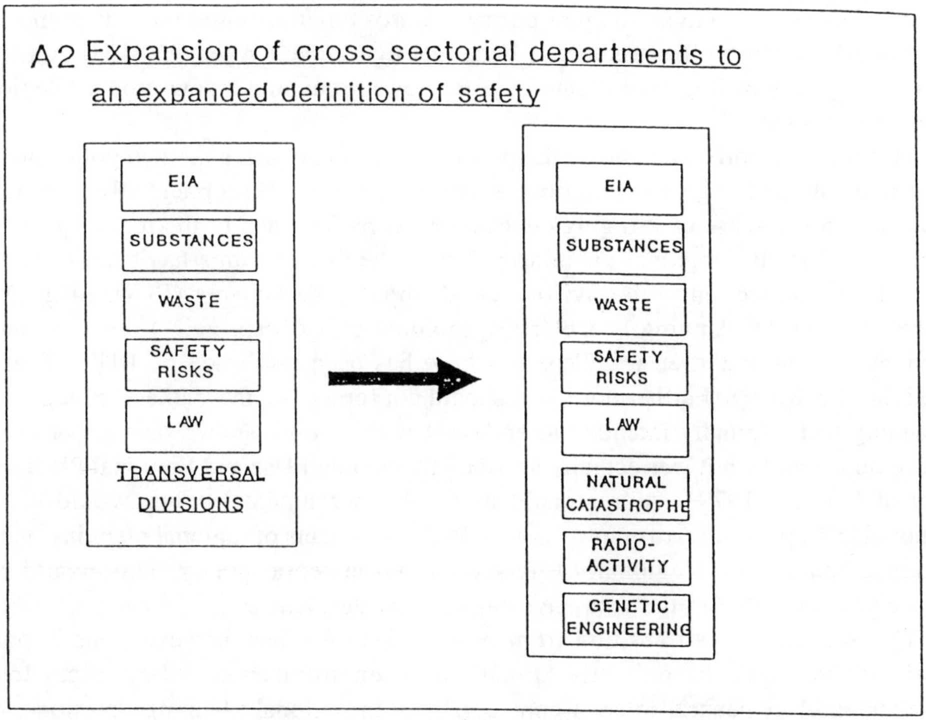

Scenario A2: New cross sectorial departments according to an expanded definition of safety.

The strict administrative separation of classical immission protection and radiation protection is nowadays hardly justified and can, at best, only be explained historically. Furthermore, risk prevention should be extended to genetic engineering. Considering the growing importance of this field, a new transversal division, 'Genetic Engineering', should be created from the current section 'Paths of substances and Environmentally Hazardous Organisms' which is part of the division 'Substances and Soil Protection'.

In the sense of concentrating powers, but also with respect to their relevance for global climatic changes, it should be investigated whether a new division 'Natural Catastrophes and Climatic Changes' should also be added. Similar considerations have presumably led to the chapter 'Climatic changes and natural catastrophes' of National Research Program Nr. 31 which was passed by the Federal Executive Council in the summer of 1990. Related functions are currently located in the Federal Water Resources Agency and the Federal Agency of Civil Defense.

4.2. *Six Scenarios for Strengthening Inter-Policy Cooperation*

Scenario B1 is the first of those scenarios that take the step from the purely protective functions of environmental policy towards environmental uses. It proposes a bridge between the (restructured) Federal Environment Agency (BFU) and the Federal Agency of Regional Planning (BRP) by introducing a new group, 'Region and Environment'.

This scenario corresponds with the political intention to bring the two policy areas environment and regional planning – often operating disconnectedly – closer together, in the sense of bringing environmental policy back to the region or, vice versa, 'ecologizing regional planning policy'. The need for this has been stressed already in connection with environmental impact assessments (EIA), local and cantonal landuse planning and planning guidelines, or (especially) noise protection. As for the latter, an excellent brochure has been published by BUWAL and BRP in March 1988. Furthermore, one should not forget that the 'fathers' of regional planning had originally intended to endow it with a considerably stronger ecological component than it was actually ascribed by the Land Use and Regional Planning Act of June 22, 1979.[21] A 'regionalization' of environmental policy would make it possible, in particular, to employ time-tested instruments of regional planning (e.g., landuse zones) in a 'regionally effective' environmental policy. This would be especially favorable in the area environment and agriculture.

This scenario for a transition from protection to use has, however, one important disadvantage: the protective functions of environmental policy might lose impetus *na-à-ns* general interest considerations in regional planning decisions, as has been the case in EIA procedures. As a matter of fact, the current practice of

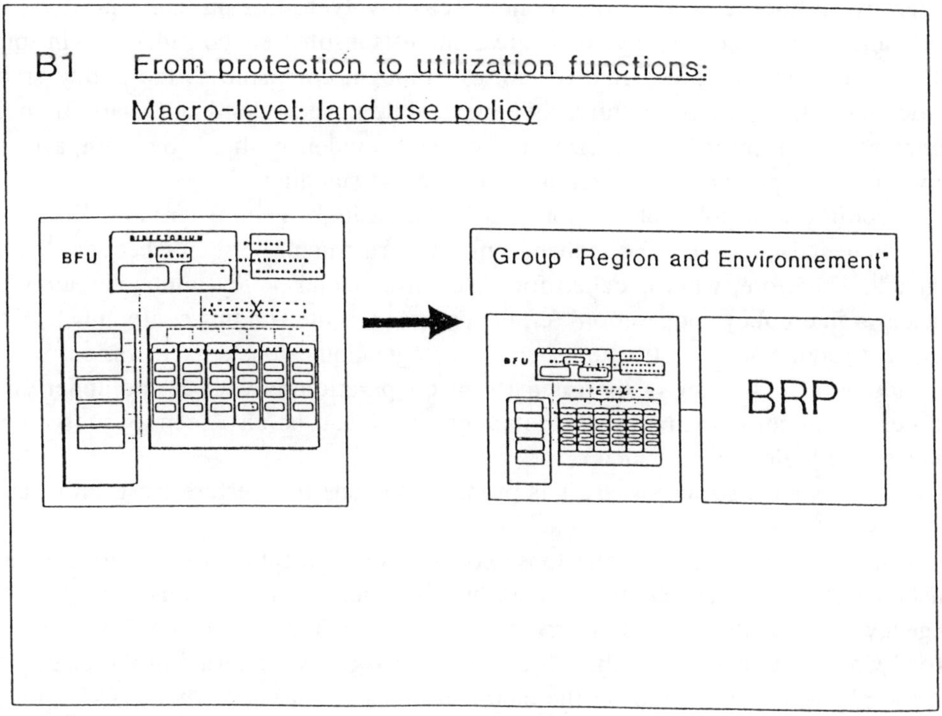

Scenario B1: From protection to use: Macro-level: Integration of land use planning.

interest consideration in regional planning does not at all help to dissipate such apprehensions, all the more so, because regional planning often fails to set definite priorities among competing user interests. Since environmental policy defines absolute positions that may not be fallen short of (e.g., emission standards), regional planning, in the case of such a 'joint venture', would have to be endowed with clear-cut patterns of privilege. As for the model presented here, it should be pointed out that the relevant federal law allows the formation of groups only if the agencies concerned belong to one and the same department.[22] Therefore, the Federal Executive Council would, first of all, have to amend the Ordinance on the Tasks of Departments, Groups, and Agencies of May 9, 1979[23] and then draft a corresponding bill to be voted on by the Federal Assembly.[24] (This vote would be generally binding and not subject to a referendum.) The same applies for scenario B2.2.

The four scenarios subsumed under B2 take yet a further step on the road from protection to use. As everybody knows, regional planning merely creates the conditions for the implementation of many other policies but plays only a marginal role in formulating these policies. That is why an even greater utilization of regional planning for environmental-political purposes in itself would hardly allow for taking a decisive influence on the other policy areas involved. Infrastructure policies, in particular, will continue to give ecological interests insufficient consideration in spite of an ecologization of regional planning, although, compared to the policy areas trade and industry or agriculture, they meanwhile have 'a dominant share in environmental disruption'. Furthermore, there is evidence that, 'by comparison, environmental policy in this area has the weakest standing'.[25]

According to a political opinion that is increasingly gaining ground. Effective environmental policy can be realized only if infrastructure policies become 'ecologized'. Therefore, what is called for is 'environmental policy implementation by means of interpolicy cooperation' (cf. the papers presented on this conference). With respect to administrative structure such an ecologization can be done in various ways. Scenarios B2.1-4 differ in the intensity of cooperation as well as the underlying strategic estimation of the present situation. They can be also seen as subsequent stages of a logical development.

Of the four B2-scenarios, B2.1 is probably the one to interfere least with traditional structures.

This scenario is based on the loose cooperation of a future BFU with agencies that are particularly relevant for the environment, such as the (recommended) Federal Agency of Agriculture and Forestry, the Federal Agency of Transport, the Federal Road Construction Agency, the Federal Energy Agency, etc. Taking the examples of the planned work group on the cantonal plans of measures or the CO_2 group, the Federal Executive Council could[26] establish a permanent interdepartmental coordinative section, 'Environmental Policy'. But more intensive and politically more influential forms of cooperation are also conceivable: The Federal Executive Council could, for example, appoint a delegation, 'Environment', consisting of the undersecretaries the federal departments of the interior, traffic, energy and economy or the general Secretaries' Conference[27] could deal regularly with environmental issues and pass the appropriate coordinative decisions.

Scenario B2.2 is based on a conception that is, as yet, quite unusual in administrative practice. It provides for a rotation of cooperating groups that are temporarily formed by different environmentally relevant agencies from various departments and the environment agency.

Underlying this scenario is the political-strategic consideration that environmentally relevant economic and infrastructure policies at the federal level should be ecologized step by step. A simultaneous ecologization of all agencies involved would be unrealistic not only for technical reasons and reasons of the substance-matter, but also for the broad political opposition this would evoke.

With this strategy it would also be possible to annul the recent transfer of forestry to BUWAL as recommended in our proposal for the agency's reorganization. The individual groups, 'Environment plus NN', to be formed should be working for at least four years. During a four-year period it would be possible to develop appro-

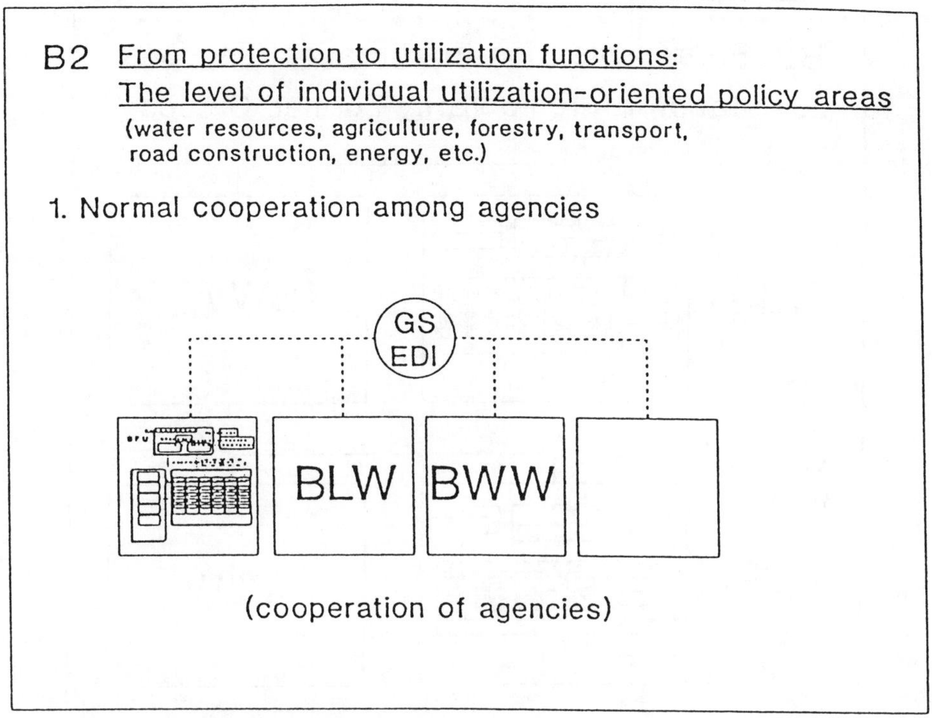

GSDEI = General Secretary of the Department of the Interior
BFU = (a restructured) Federal Environment Agency
BLW = Federal Agency of Agriculture
BWW = Federal Water Resources Agency

Scenario B2: From protection to use: The level of individual use-oriented policy areas (water resources, agriculture, forestry, transport, road construction, energy, etc.).

priate means and ways for a consideration of ecological interests both at the level of strategic legislation and at the tactical-operational level. Presumably, this will, in any case, have to be done separately for each policy area. (The principle of rotation, in general, should be applied much more widely at the level of individual federal officials too.)

The realization of the rotation model would of course entail all kinds of activities that would be though of as disturbing the 'peaceful' routine within the agencies involved. Therefore, considerable opposition can be expected to grow not only among the civil servants affected by rotation but also among the clientele of these agencies, a clientele that often is politically well-organized (e.g. Swiss Touring Club, Farmers' Association, Swiss water and energy suppliers, etc.). It can be taken for

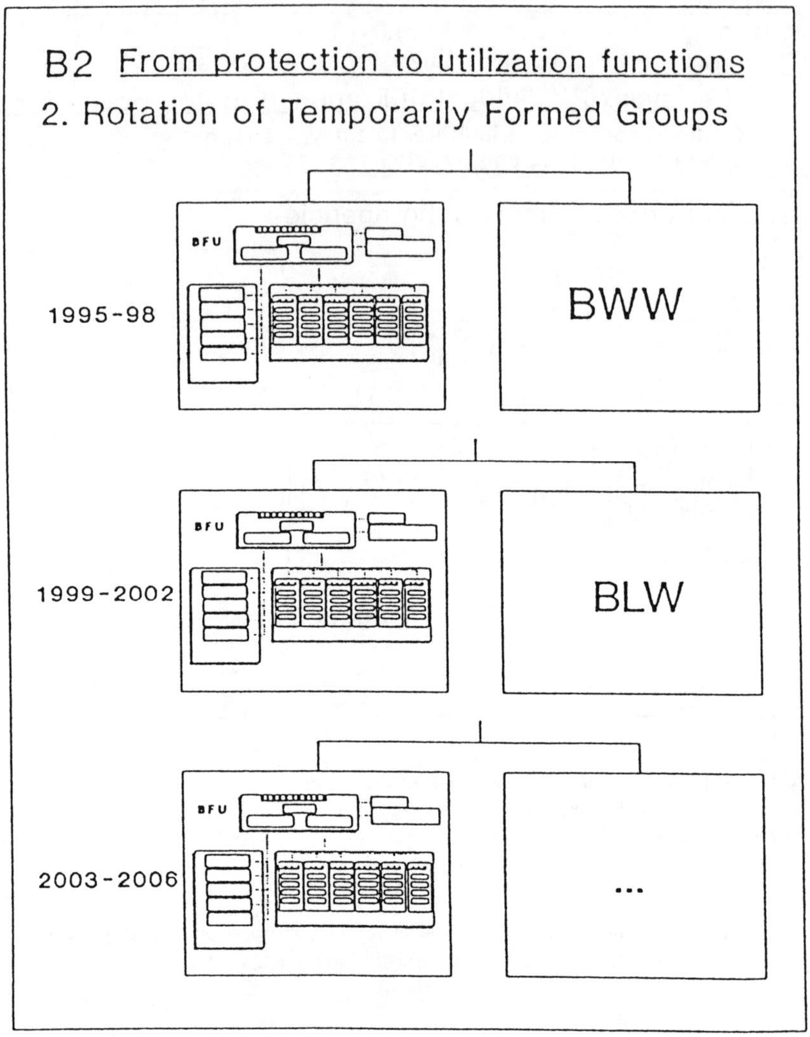

BWW = Federal Water Resources Agency
BLW = Federal Agency of Agriculture

Scenario B2: From protection to use.

granted that it will not be possible to ecologize the policy areas concerned against the open opposition of these interest groups. Consequently, it would be advisable to create structures for an effective cooperation among the administration, environmentalist organizations and these groups in order to facilitate processes of ecological learning.

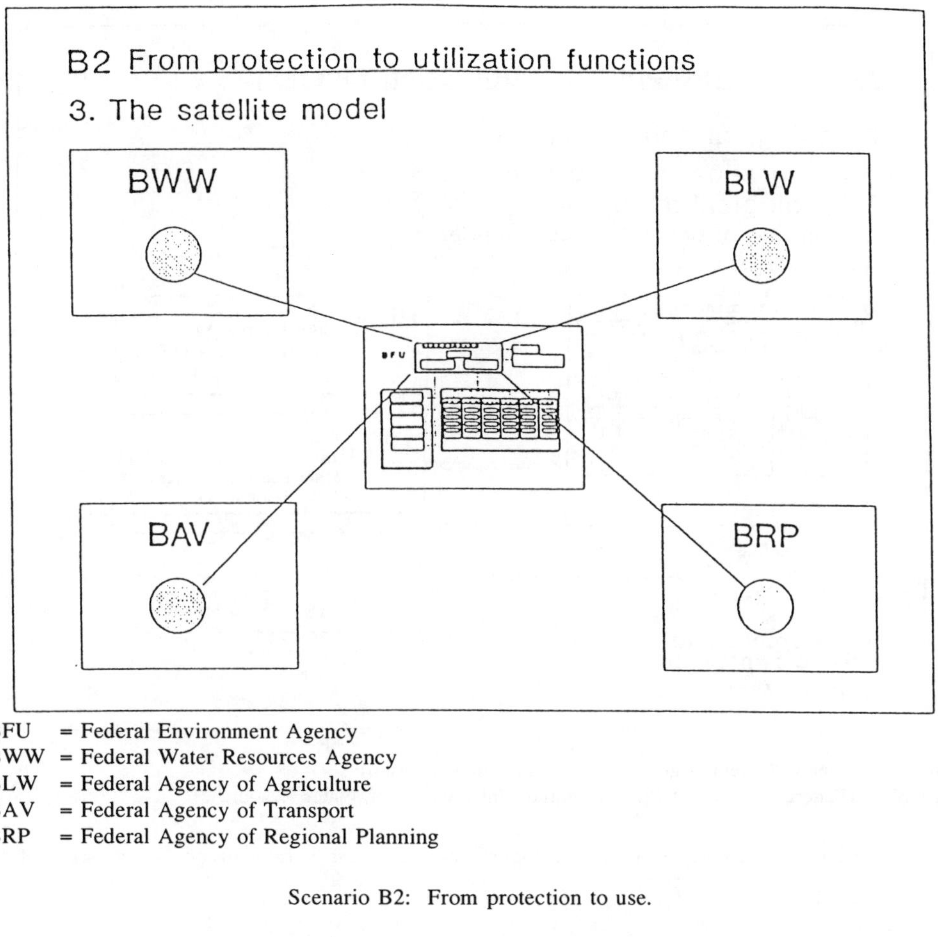

BFU = Federal Environment Agency
BWW = Federal Water Resources Agency
BLW = Federal Agency of Agriculture
BAV = Federal Agency of Transport
BRP = Federal Agency of Regional Planning

Scenario B2: From protection to use.

The satellite model shown in scenario B2.3 is based on the assumption that ecological strategies should be developed primarily within the affected agencies and policy areas themselves.

According to this model, and in adopting the principle of 'help for (ecological) self-help', the Federal Environment Agency would implant an 'ecological cell' in other agencies which would act as a catalyst and accompany the agency's ecological reorientation. Between the first B2 scenarios and this model there is a marked difference in the underlying strategic assessment of the situation (which assessment should actually be applied is, again, a matter of political decision): While the first two B2 scenarios started from the assumption that the affected policy areas themselves are hardly capable of ecological reorientation and must, therefore, be put under more or less massive external pressure, the satellite model is based on the opposite assumption, namely, that all policy areas sooner or later will perform their ecological turn-abouts in their own way. This judgment is based on the hypoth-

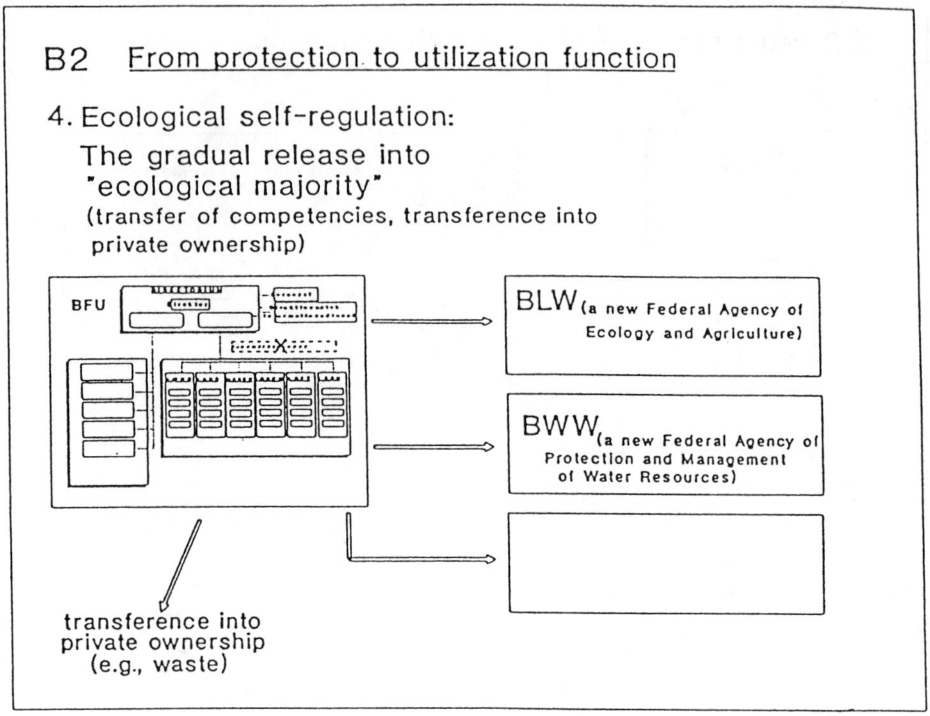

BLW = (a new) Federal Agency of Ecology and Agriculture
BWW = (Federal Agency of Protection and Management of Water Resources)

Scenario B2: From protection to use: 'Ecological self-regulation' is futuristic in the true sense of the word.

esis that at some point in time each policy area will face a situation of scarcity (scarce funds for tackling surplus production in agriculture; scarcity of land in infrastructure policy; scarcity of consensus in, e.g., transport policy, etc.). Managing decreasing resources, it is argued, forces the setting of priorities and development of plans for necessary cut-backs. Ecological arguments might at first be used only formally to seek acceptance for unpopular measures, but in the long run ecological criteria would become an integral element of planning processes themselves ('economical management of resources').

Whoever sides with this political assessment of the situation will probably be also in favor of the satellite model. And, by the way, he or she will even be able to point out first approaches to its realization within the federal administration (e.g., the 'Section Ecology' within the division of agricultural policy in the Federal Agency of Agriculture; the division 'Law and Environment' of the Federal Agency of Civil Aviation or the 'Environment and Research Office' within the division 'Projecting' of the Federal Road Construction Agency). There is, however, reason for doubt that these examples could actually prove the effectiveness of this strategy of ecologization, at least when considering our findings regarding the Federal Road Construction Agency.

As things stand, all these approaches lack one important element: The satellite model requires, at least, the institutionalized regular exchange between the environment agency and the ecological cells in other agencies, unless these cells remain hierarchically subordinate to the environment agency (which is quite unusual in federal administration). It is further required that the 'satellite' has sufficient influence on the individual agency's policy. This is generally given when it attains the status of a (though small) division with direct access to the direction.

The scenario 'ecological self-regulation' is a truly futuristic scenario. It is a logical elaboration of the rotation or satellite models in a long-term perspective. It would not make sense to take a binding strategic decision in favor of this scenario today, since the conditions for a gradual discharge of certain emission groups or certain public policies into 'ecological majority' are fulfilled neither at present nor in the immediate future. In a certain sense, the scenario represents the inverse image of the current situation. The principle of concentration on protective functions postulated above for the current structure has been dropped altogether. Exactly these protective functions should successively be outplaced from the agency and be transferred to the utilization side.

The ecology would not only be postulate at the formal or legislation level, but also be secured at the administrative-structural level. It would then fall into the competence of each federal agency respectively become an objective for the entire market economy. The result would be the emergence of a self-sustaining waste economy ('quartiary sector') or the complete integration of environmental costs in the prices at the raw material markets. The same would also apply to the various public policies. The key task of the Federal Office for Agriculture would then be ecological resources management according to clearly defined protection targets. The existing Federal Office of Water Economy would be turned into the Federal Office of Water Economy and Water Protection, and so on. It has to be observed that with the integration of the Federal Forest Administration into the FOEFL such a step has been taken (hardly deliberately) albeit into the opposite direction. The protection and utilization function have been subordinated (exactly as in the former Federal Office) to the same administrative unit. It is our recommendation that today this fusion should be abandoned. Today, the time for such a construction is not yet ripe neither in forestry policy nor in other realms of environmental policy. Our empirical findings on the (not further scrutinized) forestry policy, which traditionally incorporates strong protective functions, demonstrate, that the later are discounted again and again in favor of the utilization function.

The introduction of economic instruments in environmental policy has been

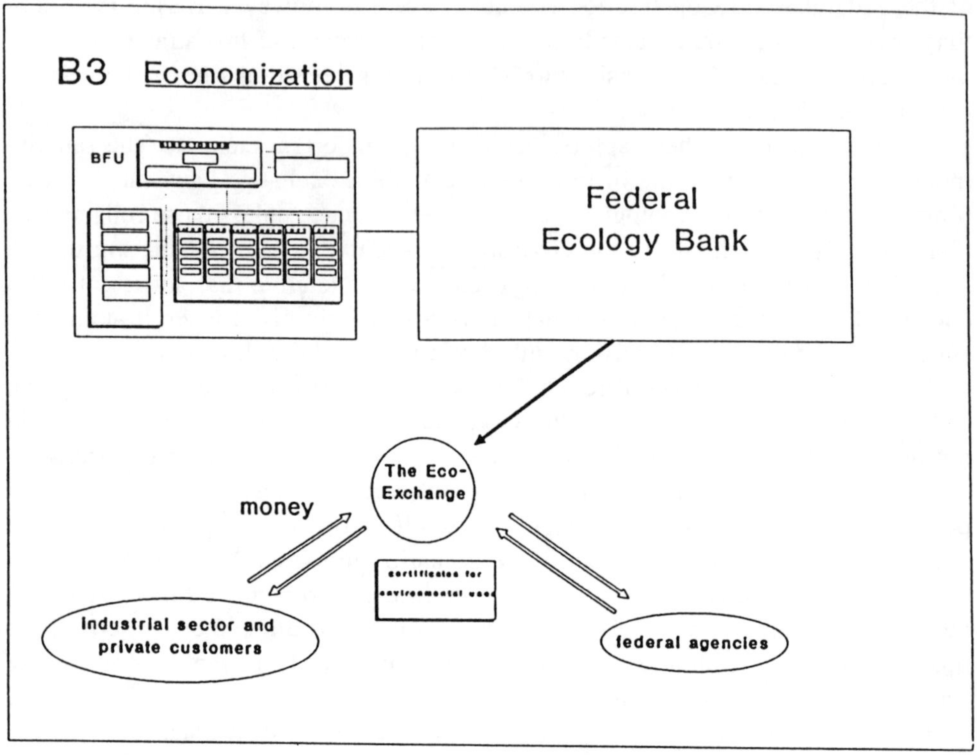

Scenario B3: Economization.

widely discussed for quite some time. Unfortunately, the various proposals hardly ever consider that such a political reform would require a corresponding administrative reorganization. As yet, it would be impossible for both the present BUWAL and a future Federal Environment Agency even to administrate the charges on solvents (VOC), fertilizers and fuel oil to be introduced by the amended Environmental Protection Act. The same applies to the so-called ecobonus, the CO_2 charge and many other environmental-related fees and charges that cantonal governments demand to be introduced.

One way to solve these administrative problems could be to entrust the federal fiscal authorities with the collection of such charges, but as for the ecologically appropriate investment of these funds, the foundation of a 'federal ecology bank' would soon become inevitable. This bank would have to ensure that funds are invested in projects that are both ecologically justified and economically profitable. Its institutionalization would become all the more necessary as soon as so-called marketable emission certificates are issued by the national and cantonal governments as a means to finance environmental protection measures that are not required by law. In that case, a federal ecology bank would also have to serve as the central exchange for such certificates.

Finally, those who politically side with an even more radical innovation and, in tendency, would like to introduce for all kinds of environmental exploitations prices that would respond to fluctuations in supply and demand, could not do so without a federal eco-bank. The introduction of marketable certificates for air pollution by industries or water pollution by households and water purification plants is frequently discussed. However, it is rarely considered in political debates to extend this concept of infrastructure policies as well, although there seems to be no convincing argument that these policies would not pay for their environmentally detrimental activities. A consistent economization of environmental policy with market-economy means must include this side too.

The introduction of such a system would create an enormous market for certificates which, just as the money market, needs to be supported by a solid administrative structure. This wold also involve a certain degree of autonomy (similar to the central bank), not only because governmental departments would also participate in this market but, moreover, in order to ensure that environmental policy can intervene in this market. If, for example, prices for emission certificates threaten to drop unreasonably, the environmental agency must be able to buy certificates in order to stabilize prices just as is routinely done on the money sector. The repurchase of certificates by the state would then not only reduce the total amount of marketable emission certificates but the general pollution load as well.

5. Comments on the Scenarios Regarding their Political Implications and their Contribution to Improving the Instruments of Environmental Policy

Preliminary Remarks

The eight scenarios included in the report on behalf of the Control Commission of the Swiss National Council do certainly not represent all feasible variants. Nevertheless, in political, organizational and instrumental terms, they cover a very broad range of possible options. In the following part, the political options behind each scenario are briefly described – a description which deliberately is not included in the report; in addition, it will be demonstrated for each scenario to what extent it facilitates a possible enlargement of the arsenal of political-administrative instruments of environmental policy. By that we understand, according to the shared definition on which this book is based, instruments with which mainly state agencies can influence the availability of resources to those target groups which are relevant to environmental policy.[28] These resources are 'money, goods, legal rights, skilled people, time, information and trust'. Since according to our knowledge there are no empirical findings or theoretical designs regarding this topic available in the existing literature, the following arguments are limited to the speculative realm of an essay, whose background consists mainly of Swiss and Middle European realities. For this reason, we will not refer to political or administrative science literature which would only provide a very general guideline for our specific topic.

Scenario 1A: New Sectorial division: Strengthening the Protective Functions

Without doubt this scenario is based on the intention to anchor environmental policy as an ecocentric policy by extending the object of protection from certain groups of human beings to all ecosystems. The resulting integral point of view is likely to strengthen the classical protection of nature and landscapes and thus the protection of rural areas as the location of a series of protectable ecosystems crucial for the regeneration of resources. The appointment of outspoken advocates of nature, probably in conjunction with the recognition of some fundamental rights for non-human, natural beings,[29] tendentiously could lead to a radicalization of environmental policy. As known, man is one of the most resistant beings, thus a relatively low level of protection is sufficient. Given the demand for an integral, ecocentric environmental policy, as put forward by the proponents of a strengthened environmental policy, the standards of protection have to be increased at the level of specific norms. One can thus refer to a vertical quantum leap in the sense of raising the standard of protection.[30]

This scenario would lead to a less compromising position on the side of the environmental agencies, especially towards the governmental infrastructure authorities, which increasingly operate large systems close to nature in less populated hinterlands of civilization. A rise in conflict and a partial break-down of currently established inter-policy cooperation relationships can be expected. The higher standard of protection could only be achieved at the price of polarization within the administration, but also in politics. In its most extreme version, the scenario would certainly be accused of 'ecodictatorship', which, regardless of its justification, would involve societal security risks having repercussions far beyond environmental issues as such.

At the instrumental level, in this scenario, environmental policy would certainly arrive at a stricter level of protection and correspondingly at lower (more rigorous) standards of immission, which in turn would result in a tightening of the corresponding standards of emission. Consequent administrative implementation would lead to a much higher importance of environmental impact assessment than is the case today, since in addition to the impacts on the environmental media air and water quality (which are typical for an anthropocentric point of view), the implications for natural ecosystems or landscapes and for established interaction systems between man and environment would also have to be examined,[31] which would again require a significant intensification of effort. In such a process, advocates of nature would have to put an (irreversible?) veto on in each case, in which even one – important – species of the ecosystem under protection is endangered. In such a conception, environmental impact assessment would become a more conflictive, institutional mechanism for the accommodation of the interests of 'minority' natural beings. The result would be a certain hierarchization of the environmental agencies at the level of central government, but also at the level of the implementing regions. A corresponding politization of science and 'nature' could hardly be avoided.[32] Instrumental gains of environmental policy (stricter environmental standards, stronger position of sensitive ecosystems in environmental impact test procedures) could thus turn into a pyrrhic victory, should the child (stricter environmental requirements) be poured out with the water (environmental policy as such).

Scenario 2A: New Cross Sectorial Departments

Such an extension of the object under protection in the light of newly emerging environmental risks corresponds to the historical development of the European environmental policies, which started with the protection of waters and were extended in the 1960s and 1970s to include clean air, chemicals and protection against technical risks. Politically such an extension means an expansion of the protection policy, which increases the importance of environmental policy, but does not imply a higher standard of protection. This can be described as a horizontal quantum leap, which in the case of nuclear safety and gene technology extends into areas which have traditionally been covered by their own regulating authorities of energy, medical and health policies respectively. The example of Germany, where in 1987 nuclear safety policy was integrated into the newly created Ministry of the Environment, demonstrates that such an expansion of environmental policy is by all means feasible.[33] The assignment of gene technology regulation to the competency of environmental policy does not necessarily reflect a radical position on ecological issues. In both cases, by the way, the (economic) utilization policies have been allocated to other agencies.

At the instrumental level, through this rearrangement, environmental policy gains the capacity to control new essential risk areas, over which it has exerted only indirect influence so far. Both in nuclear and in gene technology, a broad field of intervention possibilities opens up, through which – not at least in the interest of the respective disposal problems – it can exert influence on emerging risks at an early stage. The Sisyphus work at the waste front (including that of the two technologies) can thus be somewhat relieved. Moreover, both areas provide environmental policy access to the instruments of research and development policy and technology assessment, which goes beyond the normal environmental impact assessment and into which more ecological aspects can be incorporated. The future will show to what extent the price of environmental policy, especially in these two high risk areas, will be under pressure from the enormous economic significance of these two technologies. It will be vital to the protective function and probably issue certificates of ecological compatibility with the new ecological instruments for ecologically highly doubtful activities.

Scenario B1: Integration of Land Use Policy

This 'specialisation' scenario implicitly starts from a concept of an increased decentralization and regionalization, because it is mainly concerned about spatially differentiated goals and locally adapted protection means to be implemented, e.g. by a variety of zoning regulations, etc. Inspired by the principle of subsidiarity, which is currently being increasingly discussed within the EC,[34] the scenario assumes that environmental quality cannot be defined by the center to the last detail, since it has to be elaborated according to local circumstances and with the participation of the population concerned in its entirety. This scenario thus also represents a departure from the schematism of zoning, and it necessarily leads to a regionally differentiated policy of environmental standards. With the recognition of different climatic, topographic, but also cultural and historical parameters of environmental policy, different policy objectives result which vary according to the respective frame of reference. Politically, this scenario corresponds to a position represented in the 1970s and 1980s by the conservative center. In the 1990s, this view is gaining support both among the former rather centralist left as well as the national-conservative right. In its extreme version, the scenario could result in the abandonment through decentralization of the currently European-wide practised definition of specific standards of environmental quality at the nation-state level. In the case of this extreme version, the problem of potentially high regional disparities regarding guaranteed minimal standards of living quality can be illustrated, and can only be avoided with a – probably reduced – set of quality standards defined by the central government.

As discussed in Chapter 2, instrumental access of environmental policy to land use would be quite remarkable. Along the way, it should be possible to regulate space-critical activities of all kinds through corresponding space-specific regulations including the aspect of their impact on the environment. According to the varying degree of ecological sensitivity of the soils, various levels of intensity of the agricultural zone (maximum ceilings of fertilizer or livestock utilization), various air or water pollutant volumes per area unit in industrial or settlement zones or different waste regimes could be defined through communal utilization planing. These ceilings could be easier to implement than individual directives regarding all emittents which are active within a given area.

Moreover, such zoning regulations of environmental quality could provide more freedom for agreements and bargaining processes within the concerned collective of emittents in each zone.[35] Finally, such an approach facilitates the recognition of varying sensitivities of the ecosystems concerned through the gradation of zone-specific quality goals (probably in the framework of quality targets set by the central government). In contrast to the current widely exercised practice of defining zones which are homogeneous in regard to their utilization (living zone, industrial zone, business zone, etc.), such a regime would also allow the realization of the postulate for more inter-mixing of utilization within the respective zones, which is increasingly considered as necessary to achieve the goals of protection defined in qualitative and ecological terms.

Describing the scenario, we have already referred to the potential ecological costs of borrowing instruments from zoning policy. It would have to be taken into account that the environmental aspects also have a procedural priority over other areas of utilization, which would clearly privilege the protective aspects over the economic, infrastructure or consumptive interests of utilization in the balancing process of zone planning.[36] Otherwise there is a danger that they will be caught underneath the wheels of a philosophy of balancing interests in land use planning which treats public protection and utilization interests on an equal basis in solving conflicts over utilization.

Scenario B2.1: Intensification of Normal Cooperation amongst Federal Offices (Traditional Pattern)

This scenario is hardly spectacular. Essentially, it is committed to the current basic structure of separating protection and utilization functions based on the division of tasks among various federal offices, which in case of conflict should be accommodated not merely on an administrative, but on an institutionalized, political basis. The scenario is thus suited to render transparent conflicts between protection and utilization, to facilitate their political solution and to avoid covering them up through anticipated administrative compromises. In this respect, it would give environmental policy a politically more visible status. This has been advocated int he light of the growing importance of this policy. The transition to such a scenario would not be without consequences for the functioning of the central and regional governments, since there will be an increase of basic conflicts over environmental policy. Along this route, there will be strengthening of the environmental portfolio within governments; at the same time, the number of unanimous government decisions will decrease in favour of majority votes.

Compared with the other scenarios discussed in this chapter, the cooperation scenario is characterized by its immediate implications for the functioning of the political executive. Inter-policy cooperation is not achieved through changing the contents of the policies concerned, but through (largely policy unrelated) governmental mediation of conflicts between competing policies, which by themselves will not be resolved systematically. This scenario stands thus in clear contrast to the other versions discussed in category B, which are actually based on such an internal transformation of the essence of the policy areas concerned. It can thus be truly circumscribed as a scenario involving rehabilitating 'traditional political parties, whose influence especially in environmental policy is dwindling. Administrations, unions and interest groups of all kinds, which currently profit from a rather 'apolitical' functioning of environmental policies, are likely to have difficulties with this scenario.

This scenario cannot be described as providing an added instrumental gain for environmental policy. Replacing administration through political (non-)decisions will change the legitimacy and probably also the quality of the use of a particular environmental policy instrument (for instance, a construction permit for an environmentally disputed project). Yet the set of instruments of environmental policy as such and its relationship with other policy instruments will not be changed in a fundamental way. It will however depend on the political correlation of forces, to what extent the influence of environmental policy on the utilization of these instruments, which according to this scenario will take place in the public political realm, will strengthen their position.

Scenario B2.2: Agency Groups Based on the Rotation Principle

The rotation model contradicts an important element of bureaucratic tradition according to which once established spheres of activity of bureaucratic units can only be restricted by new political actors (ministers), and not through subministerial cooperation structures. Yet in parliamentary systems of government, shifts of environmental agencies between different ministries occur frequently. These agencies thus fall into varying ministerial contexts, in which cooperative relationships with a changing set of regulatory instances emerge. In most cases, these cooperative relationships are however looser than in those which result in a formation of groups between two or more federal agencies, as in the example of Swiss legislation referred to above.[37] The joint bureaucratic-strategic decision-making center, which characterizes the agencies linked into groups by definition, facilitates a relatively far-reaching and substantial interlocking of the joined public policies, including the use of their instruments.

Rotation is only likely to emerge under strong external (parliamentary and/or governmental) pressure, since it would lead by definition to – disturbing – processes of inter- and delocking within a relatively brief period of time ('docking-on and -off' of the respective non-environmental policies and their offices at the level of legislation, administrative structures and decision-making processes). This rotation creates dynamism, openness, substantial potential for mutual learning processes, but also bureaucratic uncertainty, which is tied to the risk that the 'wrong' partnership might have been chosen. Similar to the scenario of consequent inter-agency cooperation, the rotation scenario can only be implemented through political decision. It presupposes the capability to reach a consensus over areas of environmental priority and thus tendentiously contradicts the principle (which, in Switzerland for instance, is widely applied) that in environmental policy all target groups concerned have to make an equal contribution ('symmetry of sacrifices').[38] Popular parties which focus on large electoral segments are thus unlikely to support the principle of rotation. The model requires long-term planning periods, which go beyond the four year electoral cycles. This presupposes a certain political stability, which in our view in the 1990s exists less than ever.

The instrumental gains of environmental policy will vary according to the affiliated non-environmental policy. The potentially available instruments have been briefly described in Chapter 2 in the areas of land use, transportation and agriculture policy. Beyond these possibilities of formal access to the traditional instruments of the policies integrated in the new group, the rotation model will also fundamentally change the spirit of the administrative units which participate in the application of these instruments. Group building allows the opportunity for rotating officials among the original environmental agency and the non-environmental administrations affiliated during the four year period. In addition, new inter-policy networks will emerge at all levels, which start from shared specific problems and develop into capacities for synthesis, which will persist even after delinkage. Moreover, in a unique way the rotation scenario facilitates the concerted application of several instruments of different control modi (regulative, incentive, persuasive, etc.) in regard to one and the same objective towards a target group common to all inter-linked policy areas. The newly created networks form an almost ideal milieu of innovation,[39] in which actual explorative laboratories for developing new instruments could emerge.

Scenario B2.3: Satellite Model

As discussed above, this model starts from the assumption that a stronger presence of environmental policy is necessary in non-environmental policy areas. In contrast to scenario B2.4. ('self-regulation'), this scenario is motivated by the political belief that these policies are (at least at the present moment) not able to be ecologised.

Therefore, environmental policy needs its own representatives within the agencies concerned, who will take a certain position in the decision-making process of these policies. This position varies according to the political point of view of the proponents of this scenario (co-consultation, co-decision, veto). Though hardly openly addressed, in the event of proper blocking power being given to the environmental bureaucracy (veto position) this scenario could lead to a supremacy of the environmental agencies over the non-environmental actors of state policy. In this case, a new superbureaucracy would be created. Instead of a political-decisional, a bureaucratic-administrative supremacy of environmental policy over non-environmental policies would result.

This scenario thus has obvious features of a technocratic-bureaucratic political philosophy which is gaining sympathy among the scientists and scientific-professional civil servants who are becoming increasingly influential. The apparent decrease in the political aspects of politics is met by the growing weight of professionalism, the increasing role of science and bureaucratization. The implementation of this scenario will face substantial resistance from the concerned bureaucracies, and also from interest groups. Each administration has mechanisms to render dislike cells in its interior immune. They range from the structural isolation of satellite branches imposed on by the outside, the keeping away of them from important dossiers to personal pressure upon the disliked 'guests'. Given this situation, the satellite branch in turn, in the interest of gaining recognition by the host, tends to cut its ties with the maternal organization and to issue 'certificates of environmental compatibility' even for those dossiers which would not have been easily approved by the environmental agency.

At the instrumental level, the potential gain for environmental policy can be similarly assessed as in the land use scenario (zoning) or the rotation scenario (for other policies). Access to the policy instruments of the host policy based on a veto position seems to many committed environmental officials to be the long hoped for, almost unlimited extension of environmental policy instruments; given the lack of political legitimacy among the target groups concerned, it will almost certainly meet opposition or scepticism among the locally active interest groups. Combined with the open resistance of the non-environmental 'host' agencies, the environmental satellites would come under strong pressure, which even the most hard-nosed environmental bureaucrat would be unable to resist over time.

Scenario B2.4: Ecological Self-Regulation

As already discussed, such self-regulating scenarios are proposed as alternatives to all those scenarios which envisage an extension of the influence of environmental agencies in terms of scenarios if intra-policy cooperation (A1 and A2), and also in terms of inter-policy scenarios as discussed in B1 (B2.1, B2.2 and B2.3). In the main, representatives of the political center share the hope or the conviction that environmental policy will sufficiently be taken into constitution today by non-environmental policies. Therefore these components can be integrated into the non-environmental policies with regard to both professional and political standards. Frequently this entails the proposition to reduce the environmental administration in favor of integrating environmental skills into these non-environmental policy areas. Often this concept is seen as a counter-proposal against an alleged quest for hegemony by environmental authorities of all kinds. Where there are no serious efforts by the non-environmental authorities on their own to compensate for the reduction of the influence of the environmental agencies,it can be assumed that the self-regulation scenario is but a pretext to curb the allegedly expansive environmental policy. More serious are proposals which, similar in form to the satellite scenario, attempt to build up substantial environmental capacities in the non-environmental policy areas. Many self-regulating scenarios are based on the philosophical conviction that a rational discourse is the best means to introduce ecological reason into these other policies. The question remains unresolved, while the appropriate moment has come to discharge these policies – gradually – into ecological independence.

Such ecological independence should be reflected at all levels in the application of the traditional instruments of the policies concerned in political everyday life. In contrast to the satellite scenario, the self-regulation scenario often assumes the opening up of the concerned policies and their implementation procedures 'from below' on the occasion of the employment of instruments. The argument is that the employment of ecological instruments can mainly be achieved through procedural guarantees assuring the inclusion of the target groups and the representatives of the environmental media concerned. Tendentiously there is a decentralization and a corresponding augmentation of the policies concerned with the resource 'consensus'. On the instrumental level, the self-regulation scenario is also characterized by a harmonic conviction that even acute conflicts can be reduced through related negotiations and compromises.

Scenario B3: Economization

Economization scenarios as discussed here turn environmental policy largely into a policy of resource exploitation according to market principles. Even in their most extreme versions, these neoliberal designs envisage relatively strong regulating frameworks through which the marketable environmental utilization rights will be limited in the interest of protecting resource and avoiding cartel-like practices on the various resources market. The environmental agency will continue to play a crucial role in defining these norms. In economic terms, the first condition aims at artificially limiting the availability of existing resources through governmental norms as long as such a shortage has not been achieved. In the framework of traditional political ideologies, these economization scenarios can be attributed to European liberalism as long as they provide no solution to the related social problems. Liberalism and market economy sour the social contradictions between those, which given their economic position can afford environmental utilization rights, and the (probably largely majority) group of persons, who given their social position, is denied access to the newly created markets. In the economization scenario, this contradiction is further accentuated, as the collective consumption of resources based on governmental (infrastructure) policies is also explicitly referred to in such markets.

Instrumentally, these economization scenarios do to provide environmental policy access to the classical economic instruments which are frequently discussed today (taxes, duties, certificates etc.). According to this pure market model, the assigning of resource utilization rights to the various groups of emittents and the transaction of these resource utilization rights would not be handled by public administration, but by the corresponding markets. Environmental policy would not only be deprived of important regulative instruments (substantial dimensions of pollutant limitation through the classical project authorization), but also of essential incentive instruments, which are usually described as 'economic instruments of environmental policy'. Such scenarios would in turn make accessible to the environmental agency an enormous investment pool, which it would assume as a permanent resource restitution and management activity. It is a political decision whether this newly created 'fourth' economic sector[40] should be funded directly by the environmental authorities through the funds of assigning utilization rights, or be privatized in some way. The main instrument of the environmental authorities would be the definition (per unit) of marketable resource utilization quotas (for instance in the form of maximal admissible emission volumes), and the competency probably assigned to the authorities to acquire utilization rights themselves in order to realize specific rehabilitation tasks. The general control function over the resource utilization markets would however be assumed by stock market authority or the cartel control commission respectively. At least for the regulator, the economic instruments environmental policy would not only fail to make a gain, but suffer a substantial decrease in instruments under these economization scenarios. This corresponds to the political will underlying these scenarios to reduce governmental influence in economic and social formative processes. Governmental policy should be replaced by – sufficiently regulated – markets.

6. Concluding Remarks

In this article, we have deliberately avoided nuances. The eight scenarios are thus not analyzed in regard to their complementarity. In political reality, combinations are however feasible. Nevertheless, on behalf of the political instances concerned, the political and instrumental implications of the various options should be clarified in their most typical form. Moreover, the rapporteur is aware that a typology in the narrow sense is not presented, but rather a list of possible scenarios derived from a political discourse. For this reason combinations among the various scenarios are feasible. We hope to have demonstrated in this article that the field of the future generation of environmental instruments has to be looked for not in environmental policy itself, but in the currently non-environmental policy areas. For their development, sufficient structural conditions have to be created first. The key lies in the strategic expansion of intra- and inter-policy cooperation strategies.

Notes

* We would like to thank Dr. Stefan Kuks and Dagmar Kollande for the translation of parts of this text from German to English as well as Enzo Matafora and Dr. Giorgio Sailer for their assistance in Chapter 4 (scenarios).

1 Amongst others D. Helm, Economic Policy Toward the Environment, Oxford (Basil-Blackwell Inc.) 1991; Th. Tietenberg, Economic Instruments for Environmental Regulation, Oxford Review of Economic Policy, vol. 6, 1990; for Switzerland: R. Mauch/R. Iten/E.U. von Weizsäcker/J. Jesinghaus, Oekologische Steuerreform – Europäische Ebene und Fallbeispiele Schweiz, Chur and Zurich, 1992 and X. Oberson, Les taxes d'orientation, diss., Genève 1989.

2 Cf. P. Knoepfel, Zum Stand des Umweltrechts in der Schweiz/A Survey on Current Environmental Law in Switzerland, Cahiers de l'IDHEAP no 65, octobre 1989, p. 25 f., 57f.

3 Cf. A. Epiney, H. Furrer, Umweltschutz nach Maastricht – Ein Europa der drei Geschwindigkeiten?, EuR 1992, no. 4.

4 See P. Knoepfel, W. Zimmermann, in cooperation with G. Sailer, E. Matafora, Evaluation des BUWAL. Expertenbericht zur Evaluation der Luftreinhaltung, des landlichen Gewasserschutzes und der UVP des Bundes. EDMZ, Bern, 1991 and Rapport de la Commission de gestion du Conseil national du 18 mai 1992, FF vol. 3, no. 29 du 28 juillet 1991, p. 1422 ff.

5 This list is drawn from the Report on the State of Land Use from 1987: Conseil fédéral: Rapport sur l'Etat et l'évolution de l'utilisatiobn du sol et de l'urbanisation en Suisse (Rapport sur l'aménagement du territoire, 1987) du 14 déc. 1987, 87.074, p. 76 ff.

6 See Rapport de la Commission Popp sur 'les paiements directs dans la politique agricole suisse', Berne, 1990 and the 7th Report on the State of Agriculture, Berne, 1991.

7 See P.Knoepfel, L'agriculture et la gestion de l'environnement: Les instruments de gestion en Suisse, Cahiers de l'IDHEAP no. 60, 1989.

8 See D. Baroni, Brunner, P. Knoepfel, P. Moor, 1992: Strassenverkehrsrecht im Lichte des Umweltrechts – Le droit de la circulation routière face à la législation de l'environnement, Bale (Helbing & Lichtenhahn), Reihe Oekologie & Gesellschaft, vol. 5, 334 p.

9 See R. Imhof, W. Zimmermann, 'Massnahmenplanung' in Switzerland: The strategic air purification plan as an instrument of environmental policy in the field of urban traffic (International Congress: Implementing environmental policies by means of interpolicy cooperation, in Crans-Montana); D. Sperling, Marketable credits for vehicle emissions in California, ibid.; this question is in the center

of the COST-Action no 618 concerning "Institution Building and Information Policies" with regard to air pollution control policies in big cities in Europe (programme "CITAIR"), decided by the COST-Council in May 1992.

10 See P. Knoepfel, U. Müller, W. Zimmermann, Vorschlag für eine Programmatik sozialwissenschaftlicher Umweltforschung in der Schweiz, BUWAL-Schriftenreihe, Bern, 1994.

11 See P. Knoepfel, La protection de l'environnement en proie aux problèmes d'acceptation et aux déficits de mise en oeuvre. Mise en oeuvre de la législation environnementale: situation actuelle à la lumière des secteurs Industrie et Artisanat, Agriculture et Politique infrastructurelle de l'Etat, Cahiers de l'IDHEAP no. 70a, novembre 1991 (version française du Cahiers de l'IDHEAP no. 70), Lausanne, 40 p.

12 Report 1991 'Evaluation des BUWAL . . .', op. cit., note 4, p. 57 ff. (direct translation).

13 Ibid. p. 96.

14 Ibid. p. 96.

15 Ibid. p. 96 f.

16 See 'Rapport de la Commission de gestion . . .', op. cit., note 4, p. xy.

17 See ibid., p. 1449.

18 Report 'Evaluation des BUWAL . . .', op. cit. note 4, p. 132 ff. (direct translation).

19 Ibid., p. 149 f. (direct translation).

20 Cf. Ordinance on the Protection from Radioactive Radiation of June 30, 1976, SR 814.50.

21 SR 700.

22 Sec. 46, par. 3, VwOG = Act on the Organization of Federal Administration.

23 SR 172.010.15.

24 According to sec. 60, par. 2, VwOG.

25 See P. Knoepfel, La protection de l'environnement en proie aux problèmes d'acceptation et aux déficits de mise en oeuvre. Mise en oeuvre de la législation environnementale: situation actuelle à la lumière des secteurs Industrie et Artisanat, Agriculture et Politique infrastructurelle de l'Etat, Cahiers de l'IDHEAP no. 70a, novembre 1991, Lausanne, pp. 12, 20.

26 On the basis of sec. 55, VwOG.

27 Institutionalized by sec. 55, VwOG.

28 See the article by J. Klok, A Classification of Instruments for Environmental Policies, in this book.

29 See P. Saladin, Probleme des langfristigen Umweltschutzes, in: Kritische Vierteljahresschrift für Gesetzgebung und Rechtswissenschaft, 1989, pp. 27–55.

30 See P. Knoepfel, M. Descloux, Valeurs limites d'immissions: choix politiques ou déterminations scientifiques?, Cahiers de l'IDHEAP no. 48, Lausanne, 1988.

31 Theoretically already required: See Office fédéral pour la protection de l'environnement, Etude de l'impact Sui l'environnement, Manuel EIE, Berne 1990, p. 26.

32 See F. Ramade, L'écologie et le décideur. Les limites de la prise en compte des données scientifiques dans les politiques de l'environnement, in: J. Theys (assisté de V. Liber et M.-P. Palacios) (ed.), Environnement, science et politique, publication réalisée avec le concours financier du Service de la rechercher du Ministère de l'environnement (SRETIE), Cahier de GERMES, no 13, Paris, 1991, p. 381 ff.

33 See R. Czada, Administrative Interessenvermittlung . . . am Beispiel der kerntechnischen Sicherheitsregulierung in den Vereinigten Grosstaaten und der Bundesrepublik Deutschland, Habilitationsschrift zur Erlangung einer Venia legendi in Politischer Wissenschaft und Verwaltungswissenschaft, Universität Konstanz, March 1992, p. 120 ff.

34 See art. 3 B within the Treaty of Maastricht as well as art. 130 R, Para 2 of the EC-Treaties. (See A. Epiney, H. Furre, op. cit., N. 3)

35 Definition of emission quotas per zone-unit (ha) as already applied in the field of agriculture (dung units per hectare).

36 The only case of such an absolute priority in the Swiss environmental law is the (constitutional) obligation to protect moorelands (popular initiative introducing Sec. 24 sexies accepted in the votation of Dec. 6, 1987).

37 The history of the French Environmental Agency, which changed its status since its creation in

1972 at least six times shows that the policy contents largely influenced by the 'Corps des ingénieurs des mines' hardly ever changed.

38 Widely applied for instance in Switzerland and Germany.

39 See e.g. M. Glantz (ed.), Societal Responses to Regional Climatic Change: Forecasting by Analogy, Boulder (West View Press), 1988 and C. Jaeger, Innovative Milieus and Environmental Awareness, in: Sociologia Internationalis, vol. 28, 2, pp. 205–216; D. Nir, Region as Social-Environmental System: An Introduction to a Systemic Regional Geography, Dordecht (Kluwer), 1990.

40 In the US the waste treatment industry occupies already 3% of the active population. In France, the closing of the frontier against solid waste importation, proposed by the Minister of the Environment at the end of August 1992, provoked serious opposition from the Ministry of Economy aiming at protecting the economically increasingly important waste disposal industry.

Environment, Science & Society

EUROPEAN SCIENCE FOUNDATION

This series arises from the work of the ESF Scientific Programme on Environment, Science and Society: an economic and institutional analysis.

1. H. Opschoor and K. Turner (eds.): *Economic Incentives and Environmental Policies: Principles and Practice.* 1994 ISBN 0-7923-2601-6
2. R. Pethig (ed.): *Valuing the Environment: Methodological and Measurement Issues.* 1994 ISBN 0-7923-2602-4
3. B. Dente (ed.): *Environmental Policy in Search of New Instruments.* 1995 ISBN 0-7923-2949-X

KLUWER ACADEMIC PUBLISHERS – DORDRECHT / BOSTON / LONDON